TRILLION
YEARS
UNIVERSE
THEORY

HOW OUR COSMOS WORKS

ED
LUKOWICH

JEPKO
PUBLISHING

Trillion Years Universe Theory
Copyright © 2014 and 2015 by Ed Lukowich
Release: Canada and U.S.A. December 2015

Website: www.trillionist.com

JEPKO

AN IMPRINT OF JEPKO PUBLISHING, CALGARY, CANADA

Non-Fiction cosmology books authored by Ed Lukowich:
Trillion Years Universe Theory, (cosmology) October, 2014
(Republished December 2015).
Trillion Theory (cosmology) published December, 2015.
Black Holes Built Our Cosmos (cosmology) July-December 2015.
Science Fiction novel authored by Ed Lukowich:
The Trillionist published 2013 under pen name Sagan Jeffries.
Publisher: Edge Science Fiction and Fantasy Publishing.

Library and Archives Canada Cataloguing in Publication
Lukowich, Ed – Trillion Years Universe Theory (2014).
ISBN: 978-0-9918408-2-3 (e-Book ISBN: 978-0-9918408-3-0)
1. Cosmology. I. Title. QB981.L83 2014 523.1 C2014-900279-3
C2014-901363-9
FIRST EDITION (M-20140715). Printed by CreateSpace.

Trillion Years Universe Theory

Trillion Years Universe Theory, short form **Trillion Theory or TT,** is a new theory depicting the origin and age of our cosmos. This theory began as the brainchild of Ed Lukowich back in 1998, with the inaugural publication occurring in 2014.

Here are this author's three original non-fiction books: *Trillion Years Universe Theory* (presents Trillion Theory with the addition of an interviewer asking pertinent questions); *Trillion Theory* (purely TT); *Black Holes Built our Cosmos* (emphasizes the vital role of black holes in TT).

What Trillion Theory does and doesn't deal with:

Trillion Theory deals solely with our cosmos, namely the planets, moons, stars, suns, solar systems, galaxies, space, energy and matter which reside within that cosmos.

TT doesn't theorize about the non-physical, such as a reason for existence, or if a person has a reincarnating soul, or a spirit's afterlife. For the author's ideas on these topics, read his futuristic novel entitled 'The Trillionist,' written under pen name Sagan Jeffries. A concept in that novel is that a soul may have experienced reincarnations thousands of times on planets over a trillion years of cosmic history.

The significance of Trillion Theory in assisting planet Earth.

Surely Earth isn't the first planet to encounter biosphere difficulties, namely a deterioration of the crust, waters and atmosphere which support life. However, in Earth's case, blame for global warming is placed on humans via extreme use of fossil fuels sending emissions into the atmosphere depleting Earth's ozone shield. Also, the cryosphere of ice, snow, and glaciers, vital for drinking water, is melting away.

Harm to Earth's biological systems is ongoing. Humans have mistreated Earth; hopefully this isn't an irreversible path, unable to support future life. There may come a time when the only recourse for survival is relocation to another planet - an option far from viable with present technology.

Certainly, humans were blessed with 'the perfect planet,' a phenomenal Goldilocks candidate, not too hot nor too cold, situated in an orbit an ideal distance from a sun. Was such perfection accidental, or did some Caretaker initially oversee a deliberate terraforming of the environment of Earth making it a prime candidate for hosting life. If so, what would this Caretaker think of the way we today treat the planet which has been placed under our care?

Earth has been hospitable to life for millions of years; a hospitality which could be shamefully destroyed during man's comparatively short history. Thing is, we don't know just how fragile the 'living quarters' of a planet such as Earth are. The main question seems to be, 'If technological pollution is ruining our planet, can a lowering of pollution alone bring about a reversal back to good fortunes? Or, does the present situation dictate that newer technology be found as a means to refurbish/revitalize the biosphere?'

Our future requires that we figure out how to fix the environmental problems we've caused to Earth. There is a need to thoroughly understand our cosmos. This is where Trillion Theory can provide hope. The development of any Earth-saving technologies may be very dependent upon a utilization of new cosmic ideas presented within TT.

Trillion Years Universe Theory

Trillion Theory
is a new cosmic rival
with the intent of upending
the Big Bang's explanation as to
the origin and age of our cosmos.
According to the Big Bang, supposedly
a **huge explosion** occurred at **13.7 billion
years** in the past spreading hot gases across
the expanse of **waiting space**. Supposedly, the
hot gases coalesced via nebular spin to form the
spheroids, solar systems, and galaxies so prevalent
throughout cosmos. Contrarily, we know explosions
spread outwards in an equidistance from center. This
means an entirety of our cosmos should be roundish,
but it isn't. Rather, TT says that our cosmos is shaped
more like a thick fat cucumber. If TT shape is correct,
then a Big Bang never occurred as the event to best
explain the origin, age and the colossal size of our
cosmos. Also, the false Big Bang has no means to
explain atoms or to answer the vast number of
pertinent questions dealing with cosmic
intricacies. Thus, Big Bang fails all the
litmus tests. It is time for TT
to displace the
Big Bang.

Ever wondered about a different cosmic explanation?
Ever wondered what powerful secret keeps
planet Earth rotating perfectly on its axis?
Ever wondered why our solar system has a sun?
Ever wondered how the humongous galaxies
of our cosmos became solar system hotels?

Trillion Theory provides answers to origin questions right from a quantum atom world all the way up to the macro world dealing with billions of galaxies, offering a far more realistic explanation than Big Bang theory as to how and when cosmic events occurred. TT varies greatly from **Big Bang's supposed huge explosion:** TT shows how our cosmos was shaped from eons of growth rather than an explosion. Instead of Bang's **13.7 billion years,** TT calculates cosmic age at over a trillion years. According to TT, every 15 billion years or so, solar systems die and recycle to reform into brand new solar systems. TT calculates 67 recycles over a trillion year history. Stars in our sky are simply the latest rendition.

Waiting space: TT suggests that space was not waiting for our cosmos to be born, rather all that existed at the origin of our cosmos was an endless ocean of energy and black holes. These black holes accessed this energy utilizing it to build matter around themselves to become cosmic spheres, leaving behind the large empty cavern we call space. As our cosmos grew with each 15 billion year recycle, more of this ocean of energy was utilized to build billions of ever-larger galaxies. TT secrets are unveiled within this book.

Trillion Years Universe Theory

TABLE OF CONTENTS   Page

FOREWORD

By Brian Hades of EDGE Science Fiction & Fantasy Publishing. Brian published Ed's futuristic sci-fi novel *The Trillionist* under Ed's pen name Sagan Jeffries.

"When Ed first approached me about supporting his new universe ideas, I was immediately intrigued. While not brandishing an astrophysics pedigree, Ed does possess an absolutely unique perspective of our universe. His new model depicting how our cosmos originated and then grew to its present gigantic size is captivating in *Trillion Years Universe Theory*. Ed takes our cosmic origins back in time a trillion years, far past any Big Bang of just 13.7 billion years,

If you regard our universe as mere happenchance, think again! Ed's theory provides a vivid description of black holes as miraculous inventions designed to be cosmic builders and organizers. His Trillion Theory is far different, radically changing our cosmic view.

Ed would ask, 'Have you ever wondered how our solar systems and all the galaxies came to be? Or, why our cosmos is so gigantic with billions of light years across its expanse?'

If you are interested in discovering a new laid out path pertaining to the origin, history, operations, and organization of our cosmos, then this is the book for you. I recommend this Trillion Years Universe Theory book as an incredible must-read for everyone."

T R I L L I O N Y E A R S

PREFACE to TYUT

Book description: Ask a pertinent question about our cosmos - TT promises a new radical answer. Ex: What is at the central core of every planet, moon, and star? The hidden secret is unveiled in TYUT's new theories.

Reason why new theory written: The author became disenchanted with past theories depicting the origin of our cosmos. A decisive factor in committing to the complex task of developing new theory was the need for disproving the old sacred Big Bang. TT genesis began back in 1998 by thinking outside the box and strategizing how a universe could have been built.

Purpose of Trillion Theory is to propose a new way to determine origin, age, and function when unlocking cosmic secrets. It is a critical step for humans to gain a mastery over their cosmos. The benefits of such an accomplishment would be immeasurable. Right now, the time is ripe to conquer the final space frontiers.

TYUT's targeted audience has an acute interest in our cosmos. Millions of adventurous persons are amazed by the findings of astronomers and they are anxious to hear of possible answers to cosmic mysteries.

What author of TT has learned: To be candid, Trillion Theory is a little sci-fi, plausible, but with a grain of salt. For the author, there was a greater appreciation of our cosmos when writing all the theory from a 'universe builders' strategic point of view.

T
R
I
L
L
I
O
N

T
H
E
O
R
Y

INTRODUCTION:

BY AUTHOR
ED LUKOWICH

'Astronomers admit they are re-thinking our universe. Large cracks are appearing in the Big Bang theory.'

Back in 1998, I was hard at work organizing my new ideas theorizing about how our cosmos had come to be. To formulate my new Trillion Theory (TT), the main license I wanted was that of being human, born with the right to search for life's answers. The failure of others to properly explain of our universe perplexed me. Also, I found the Big Bang simply too haphazard to quench my thirst.

Yet, my wheels spun. For, my TT manuscript depicting our cosmos as this ancient trillion-year-old growing entity found itself collecting dust on a book shelf. As a busy Olympic curling coach, icy time slid by. Finally, 15 years later in 2013, I took my first steps with the release of 'The Trillionist' (sci-fi novel published under my pen name Sagan Jeffries). Readers ate up my far-out ideas relating the story of an unscrupulous entity, older than dirt, which cheated regular cosmic rules.

In 2014, 'Trillion Years Universe Theory' was published as my inaugural theory book. TYUT is Trillion Theory plus many theory questions from an interviewer. The best way for me to describe Trillion Theory is as a brand new theory depicting a far different origin to our cosmos, plus the means by which our cosmos has grown via the recycling of its spheres and solar systems over the trillion year span of galactic history.

Thus in TT, older stars can exist alongside young stars, just like in a recycling growing forest where old tall trees stand tall beside small saplings. TT takes recycling out to the spheroids, solar systems, and galaxies of our cosmos.

In TT, it is helpful to think like a cosmic strategist.

Till now, no other theorist has made as strategic a theory as TT. For, TT approaches the design of our cosmos from a strategically scientific perspective. A central question which the author posed to himself at the outset was, 'If one had to strategically design and construct a cosmos to meet strict TT parameters: simplicity coupling complexity; similar themes from micro to macro; one type of basic material to supply the cosmos; one type of engine to continually mold this material into spheroids and thereafter spin and organize these spheres into solar systems and galaxies via pre-set cosmic rules; and lastly establish self-perpetuating recycling; how would one approach such an enormous task?'

If determined to undertake this gargantuan venture, then what measures would have been taken by an Artisan, or a corporation, or some computer genius, or the combination of all these and more, to invent such a unique universe?

Some experts have unknowingly helped me.

During my writing of Trillion Theory, I contended that TT solved mysteries which still continue to baffle theorists who believe in the Big Bang and Nebular theory and incorrectly estimate the age of our cosmos at only 13.7 billion years.

Over the past 17 years, each puzzle that an astronomer or astrophysicist ran into seemed to always do more to confirm the validity of my Trillion Theory. Here are a few examples:

Gemini Telescope program in 2004 found galaxies more fully mature than one might expect in a 13.7 billion year cosmos. Dr. Robert Abraham, Dept of Astronomy, U. of T, "We are seeing that a large fraction of stars in the universe are already in place when the universe was young, which should not be the case. This glimpse back in time shows pretty clearly that we need to re-think what happened."

Dr. Patrick McCarthy, Observatories Carnegie Institution, added, "It is unclear if we need to tweak the existing models or develop a new one in order to understand this finding."

Then in 2014, astronomers found in our Milky Way Galaxy a star which they named Methuselah. At 16 billion years of age it was older than the supposed 13.7 billion year origin of our cosmos. Big Bang had no way to explain Methuselah.

Whereas, my Trillion Theory model shows why older stars can exist in a supposed young part of our universe.

Is TT right, wrong, or just food for thought?

As the founder, I readily admit that TT is very radical. As a reader, your thoughts might be, 'What cockamamie stuff is this?' If so, I can only request that you keep an open mind.

Right: It'll take a prolific discovery to prove Trillion Theory, such as an astronomer's announcement, 'Yes, we witnessed the battle taking place between black holes right after a supernova destroyed its solar system.' Even partially right would be a great celebration for Trillion Theory.

Wrong: What if Trillion Theory is dead wrong? Then, I'd be the first to welcome a better proven cosmology theory.

Food for thought: What if no one can prove Trillion Theory? Then, perhaps it is merely food for thought to help expand our thinking while endeavoring to understand our cosmos.

CHAPTER 1
PREVIEW OF
TRILLION THEORY

'A brand new theory vastly expands our ideas.
No great idea was ever immediately accepted - most
were initially thought of as ridiculous.'

Trillion Theory is new cosmology theory proposed by self-proclaimed theorist Ed Lukowich. This TT theory proposes a totally different origin for our physical cosmos than proposed by the Big Bang theory.

Also, TT pays particular attention to specific roles which black holes played dating back to the origin of our cosmos. TT says that black holes are major entities involved in the growth, recycle, organization, and operation of our cosmos.

The purpose of this chapter is to provide a tasty preview to TT. As a new theory, TT is detailed, and rightfully so, for our cosmos is complex beyond belief to the extent that the normal 13 seconds needed to explain Big Bang or Nebular theories simply won't suffice. With that in mind, this preview seeks to offer a brief glimpse of new Trillion Theory with the goal of helping the reader to determine a course of action. **This TT Preview begins with enough fuel to build a fire.**

TT proclaims our cosmos to be a trillion years old, thus far older than 13.7 billion years - which TT points out as just the age of the current rendition of stars in our sky.

TT is far different than Big Bang. So, it is necessary to look at the big cosmic picture to see what TT depicts differently. To begin, we move to a unique vantage point, outside of our known universe, seeking a big picture view. From there, we easily recognize certain things we already know. However, when something new is seen, TT will be sure to point it out.

Shape of the entirety of our cosmos, according to TT.

According to Trillion Theory, the generalized shape of the entirety of the 'space' of our cosmos is not round but rather oblong, longer in direction; similar in shape to a cucumber. The reason for this elongated shape will be explained later.

The entirety of our universe's space is a cucumber shape.

From outside of this entirety of space we can see inside noticing billions of prominent galaxies.

In the space of our cosmos there exist billions of galaxies. According to TT, many of the larger galaxies are hundreds of billions of years old. TT further suggests that the multitudes of solar systems existing inside these ancient galaxies are all in various phases of their recycling process. This recycling has been going on in cosmic history for a trillion years.

Now, according to TT, the general location of most of these galaxies within cosmic space is generally away from the center, closer towards outer perimeter edges of space. The reason for this will be explained in this book.

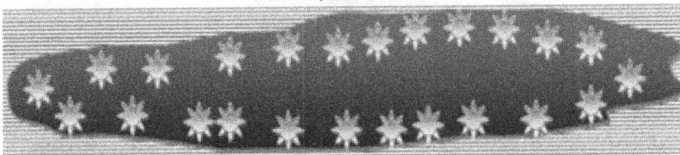

A rough model of our present universe, with billions of white-colored galaxies positioned outwards away from the center.

Outer boundaries of our cosmos.

According to Trillion Theory, the outer limit of space is not the outer limits of the entirety of our physical universe.

Astronomers consider that the outer limit boundaries of our cosmos are as far as space extends outwards. Thus, our physical universe is supposedly everything inside of this space area. If this were correct, then everything available to coalesce and form spheres is already inside of this space. Meaning that's all there is, with the amount of matter limited and constant, with no means to increase the size of cosmos.

On the contrary, according to Trillion Theory, out past the edge of space there exists a supply of yet unused material readily available to be accessed and deployed to continually increase the matter content inside the space of our cosmos. According to TT, utilization of this external supply of material has been a key feature allowing for continual cosmic growth over the past trillion year history of our physical universe.

What is outside of this entirety of space?

According to TT, there exists tons of available energy. The relevance is that our cosmos is always afforded a continual source of new energy to access and then convert into matter in erecting billions more of its solar systems, and galaxies.

Out past the outer limits of space exists a supply of material (shown in white) ready to be used to continually increase the count of solar systems and galaxies inside our cosmos.

Therein, TT extends the outer boundaries of our cosmos showing how the limits of space are continually expanding.

In the illustration, TT displays the existence of a gigantic sea of light energy surrounding the space of our universe. This ocean of light energy sits in a static frozen form readily available to be used for more building inside of our cosmos.

Energy surrounds black space. Galaxies position themselves towards the outer edges of space to access the energy.

An analogy would be a normal mining operation. Miners burrow deep into outer walls of a mine, reaping ore, while at the same time ever-extending the overall space of the mine.

In TT, as more energy is accessed and brought inwards from the outer static ocean, more solar systems and galaxies are built. As this static ocean of energy is dug into, nothing is left where energy was depleted, nothing that is except for newly created space. It is possible to build more galaxies and simultaneously extend space further afield.

Zoom inwards to see the contents of our cosmos.

Inside of our physical universe, space comprises over 99% of the area. Thus, huge empty distances of black space, void of matter, exist between the galaxies. This enormity of space seemingly makes for insurmountable travel distances?

Inside of our cosmos, matter comprises less than 1%. This matter is housed in oasis galaxy islands sitting in space. TT affirms these galaxies to be hundreds of billions of years old, far older than the supposed Big Bang's 13.7 billion years.

The most common galaxy is the spiral. There are billions of these, all seen as isolated islands, spinning clockwise or counterclockwise around a supermassive black hole.

Zoom further inwards to view one galaxy.

A spiral galaxy is a pancake shape with a central bulge. TT later examines the cosmic laws which bring about this shape.

At the hub of the spiral galaxy there exists a supermassive black hole using its powerful spin and its gravitational pull to hold the galactic contents (namely solar systems). Gigantic stars forming the inner body of the galaxy are surrounded by several swirling spiral arms. The body and arms all revolve in orbit in the direction of axial spin of the supermassive.

Do galaxies spin clockwise or counterclockwise?

Astronomers find this to be a 50-50 proposition: half of the spiral galaxies spin clockwise and half counterclockwise. Thus, cosmos is both left and right handed. This is relevant in disproving Big Bang and Nebular theory where cosmos should be one directional. Astronomers and astrophysicists have no way to explain this 50-50. Whereas, TT will be able to explain why this 50-50 cosmic phenomenon exists.

Trillion Theory (TT) takes a closer look at cosmic spin.

We will use a spiral galaxy as an example. The stars of the spiral galaxy are in orbit around a supermassive black hole which in this example pivots and spins clockwise on its axis. All of these stars revolve in clockwise orbits around the supermassive because they are tugged along by the reach of supermassive's gravity. We already know this important law about gravity: 'All objects caught and held by gravity must revolve in orbit in a direction as dictated by the direction of axial spin of the central body which is creating the gravity.'

In this example, the dictating supermassive black hole pivots clockwise on its axis creating a clockwise moving gravitational field surrounding itself. Objects, caught in this field, orbit in a clockwise direction around the central supermassive black hole.

This law pertains to galaxies such that the direction of axial spin of the central supermassive determines the orbital direction for the entire galaxy. This law also pertains to solar systems such that the direction of axial spin of a sun determines the orbital direction for the entire solar system.

Now, Trillion Theory makes an important distinction.

We already know from astronomy that, 'revolve in an orbital direction' is not the same as 'direction of pivotal spin on an axis'. Later, TT will show this distinction as paramount in explaining how a regulated TT cosmos operates.

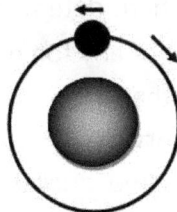

Overhead view of a planet in orbit around a sun. Here, the planet is orbiting its sun in a clockwise direction (the arrow on the right) because this particular sun pivots clockwise on its axis creating a clockwise gravity surrounding itself. However, in a contrarian motion, the planet is pivoting on its own axis with a counterclockwise spin (as show by the left arrow).

Examining solar systems TT style is the next step.

There are reportedly millions to billions of galaxies in our cosmos. Each mature galaxy can house millions of suns. TT states, 'suns with solar systems are the norm, meaning that there are billions of solar systems in our cosmos.'

In our solar system, our sun pivots counterclockwise on its axis and its massive gravity holds 8 planets in orbit, all orbiting counterclockwise. Then, each of these 8 planets possesses the gravity to hold a moon(s) in orbit. While Earth has only one moon, larger planets have many more. Over 150 moons have now been identified in our solar system.

In our solar system, all the spheres don't axial pivot in the same direction. Our Sun, and planets Mercury, Earth, Mars, Jupiter, Saturn, and Uranus all counterclockwise axial pivot. In contrast, Venus and Neptune pivot clockwise on their axis.

TT formulates a new cosmic law governing why the axial spin direction of planets and of moons is independent of the direction in which they orbit their sun and also independent of the axial spin direction of their sun. Basically, TT will show that a programmed sphere determines a private direction of axial spin, independent of any force or gravity acting upon it. No other theory other than TT can explain how.

Zoom inwards even further.

Zooming inward to the quantum micro world, we find that all the matter forming spheres is a complexity of many continuously spinning subatomic atoms. Over 100 types of atoms, each with a definitive subatomic structure, have been catalogued. Neither Big Bang nor Nebular theory can explain how or why these atoms exist. Whereas, TT can explain how atoms were made in the past and are still being made today.

Zoom partway back out to see what we have. A hotel is good analogy for galaxies and solar systems.

Imagine the galaxies of our cosmos as island hotels. TT says that a galaxy is an established permanent hotel, up to hundreds of billions of years old. Occupying the rooms of this ancient galaxy hotel are non-permanent solar systems which have a shorter lifespan (upper limit 15 billion years).

Every unit solar system is independent of every other system in that galaxy. Any one particular solar system can check out of its room when its own sun goes Supernova, without normally affecting any other solar system.

Trillion Theory will show that when a sun ages out, goes Supernova, explodes and destroys its solar system near the end of its 15 billion years, then that particular hotel room is made available for a new and larger tenant. Or in some cases, two new rooms will replace the one old room, thereby upping the number of galaxy solar systems. TT will show how this contributes to a population growth for our cosmos.

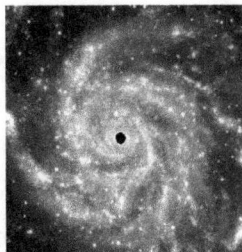

Structurally, the Hubble Space Telescope shows that each of these humongous galaxies has a supermassive black hole as its bulging central hub. During the extremely long epoch of cosmic history, billions of gigantic galaxies, far separated from one another, formed up into islands, acting like hotels. Within each galaxy, there formed multitudes of quite-large units known as *solar systems*. - acting like rooms in the island hotel.

TT's true age of various components of our cosmos.

According to TT, our cosmos began at least one trillion years ago (see chapter 5 math). Inside cosmos, TT calculates the following ages for cosmic entities: galaxies vary from 100-800 billion years; inside galaxies, the recyclers are the solar systems and their spheres at less than 15 billion years.

Where black holes enter and fit into Trillion Theory.

TT proclaims that **black holes** are prevalent throughout all the galaxies. TT sees a **supermassive black hole** at the hub of every spiral galaxy. Furthermore, **TT claims the existence of a cloaked black hole (size dependent, either xxx-large, x-large, large, medium or small size) hidden away at the core of every star, sun, planet and moon in our cosmos.**

While Big Bangers still try to fit the discovery of black holes into their thinking, TT states that our *cosmos could not exist without black holes.* They are absolutely necessary.

TT shows how black holes operate like machines, building and also bringing orderly organization to galaxies, and solar systems. Black holes have been doing their job for a trillion years, recycling the matter and energy of our cosmos.

Black holes have built billions of solar systems in the past, and our solar system is simply the latest which we know of.

Unfortunately, all the past solar systems of past ages have left no trails to sniff; no graveyard head stones to prove they ever existed. Yet, TT knows they did exist. As in the ancient past, today black holes by the trillions are busy building.

"How did those cloaked black holes get inside of stars, planets, and moons?" TT will demonstrate that a naked black hole has the process ability to spin matter from energy as it builds the body of a sphere around itself.

So to prove TT, it will be necessary for astronomers to look for naked black holes in different ways. Years ago, it was surmised that a black hole was a singularity left behind when a star went Supernova, exploded, and then finally imploded back into itself. Now, TT suggests that the black hole was at the center of that star from the time of the star's inception.

Trillion Theory (TT) presented here shows:

• How our cosmos began a trillion years ago.

• How moons/planets/stars were born.

• How spheres recycle approximately every 15 billion years.

• How cosmic spheres double each 15 billion years.

• How our recycling cosmos is of infinite design.

• Cosmic spheres being in different phases in their cycles.

• Why our entire cosmos is linear in shape, not spherical.

• How galaxies receding away from each other is an outward pulling force (not from an internal explosion).

Foundations of new Trillion Theories:

• Trillion Theory places LIGHT as the material at the forefront of our cosmos. Ever humble, light displays more of its many uncanny talents, upping its phenomenal properties. TT will demonstrate how light is incredibly spun into matter.

• TT drastically changes the role of BLACK HOLES, showing them to be unique catalytic engines which built the spheres of our cosmos by spinning light into matter.

• TT provides a better explanation of what SPACE is and how it formed; and, how TIME is a specialty imported item.

Where gravity enters the TT picture.

What really is gravity? Where does it come from? When we talk black holes, TT talks about the gravity they produce. TT proclaims that black holes provide gravity as we know it.

Throughout, gravity is the great constant associated with each and every black hole. All black holes pivot furiously on their axis which allows them to generate tremendous gravity. Naked black holes are swallowing pits for energy passing nearby, even to the extreme where they are even capable of unraveling nearby stars into early Supernova action.

The importance of various levels of gravity in TT.

Black hole size is crucial factor when TT writes cosmic laws regarding gravity. The larger a naked black hole, the greater is its power to out-battle lesser sized black holes in competition to devour light and spin that light into matter.

The larger a black hole, the greater is the mass of the sphere which it can build itself into. With larger mass, the larger black hole can project a stronger gravitational field and thereby hold more spheres in orbit around itself.

Thus, the largest black hole possessing the strongest gravity of all the spheres in a solar system will be inside of the sun which centers that solar system.

Similarly, the largest black hole possessing the strongest gravity of all of the spheres in a galaxy will become the hub supermassive black hole centering the galaxy. Supermassives are the mightiest black holes, holding multitudes of stars and their solar systems in the spiraling arms of the galaxy.

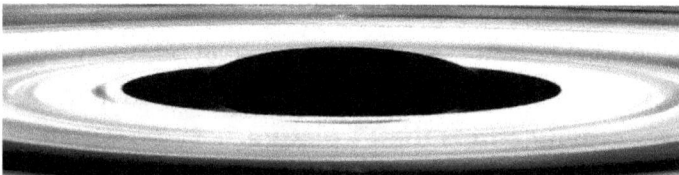

Gravity provided by the supermassive powerful black hole at the central hub of the galaxy keeps the entire galaxy (hotel) together and spinning as one.

What should the reader expect in this book? Is Trillion Theory (TT) difficult to understand?

No, not at all difficult, no crazy formulas to digest. Only formula is Einstein's $E=mc^2$. So, anyone can read new Trillion Theory and easily grasp the concepts. Agreeing with the new ideas is entirely another matter. For instance, it's easy for us to presume that galaxies and solar systems formed in space. That space was already there, ready and waiting.

However, TT sees space differently, contending that space was not there waiting for matter to appear. Rather TT says that space is a leftover byproduct, namely the emptiness left behind when matter formed into the spheres of our cosmos.

"Whoa" you say, "Space had to be prior to anything else."

But, TT sticks to its guns by redefining space. Also, TT will provide new ways to think of cosmic aspects, namely: origin; age; history; shape; size; perimeter; time; and cosmic laws pertaining to formation and orderly organization of galaxies, solar systems, stars, planets, moons, and even atoms.

Here is what you can expect to find in this TT book.

- True purpose of black holes as builders in our cosmos.
- A look at the outside around a black hole.
- A look inside of a black hole (never seen before).
- The cosmic laws by which black holes operate.
- Possible future proofs for Trillion Theory.

When it comes to our cosmos, our eyes can fool us.

Centuries prior to 1500 A.D., people's eyes fooled them. They watched the moon and stars move across the sky. It was natural to believe that they were the non-moving center and that the objects in the sky were the movers. In error, they were convinced they centered the universe.

Today, might our eyes be fooling us once again? When we look at planets and moons, do we view them as hunks of matter? TT views them as part of a cosmic recycling process, and for now they are doing a good job keeping control of the atoms from which they were made. Whereas with our sun, TT sees a sphere in its later innings, losing the struggle to control the atoms from which it was made.

TT says that those planets and moons will also one day lose control of their atoms - they have no choice. For matter always returns to energy; an energy available to be recycled by a black hole and spun into a new sphere in our cosmos.

Trillion Theory (TT) attacks the Big Bang.

When astronomers make a new discovery they often try to make it fit into how the widely accepted Big Bang model depicts the origin and age of our cosmos. Sometimes they skew results, hoping to fit a square peg into a round hole.

Instead, TT says *"no"* to Big Bang. TT presents new theory stating that our cosmos is much older than 13.7 billion years. TT calculates our cosmos at one trillion years of age when using TT's new recycling model. Recycling always blends one generation of solar systems into the next, while 15 billion years is the upper age of any one particular solar system.

History shows how the Big Bang became a theory.

Back in 1929, Edwin Hubble (Cosmic Expansion) observed that galaxies were indeed redshifting on the color spectrum, indicating they were receding in a definite cosmic expansion. Then in 1931, Georges Lemaître proposed that the redshift was due to an explosive origin. He postulated that a central singularity of matter originally exploded dense hot contents outwards to swirl and cool to form spheres.

In 1949, English astronomer Fred Hoyle in a BBC radio broadcast rejected the theory as 'cartoon physics' mockingly coining *Big Bang*, inferring to the theory as a ridiculous hoax.

The real clincher for Big Bang came in 1965. Penzies and Wilson showed the glow from 'cosmic microwave radiation' as almost exactly uniform in all directions, thereby indicating expansion. Big Bang became the accepted origin theory. Sadly, any new discoveries must fit Big Bang's stranglehold, or get twisted. Or, they are shelved into the unknown file.

Trillion Theory (TT) sets out to disprove the Big Bang

TT argues that Big Bang theory is wrong, and that there are better explanations for redshift and cosmic microwave radiation background. TT shows how 'isotropic' (namely the same in all directions from a center which should occur from a Big Bang explosion), is not the true model of our cosmos. Today, most astronomers are agreeing by predicting that the shape of our entire cosmos is more linear than spherical.

TT properly accounts for the linear shape seen in galaxies and solar systems, where the configuration for the spheres is flat and extending nearly straight outwards from a center.

TT argues that astronomers haven't looked hard enough to attempt to find another more plausible reason other than Big Bang for this outward expansion of our universe.

For instance, Big Bang (as the explosion advocate) bitterly fails to explain spin. So, it recruited Nebular Theory to assist. Supposedly, gas-cloud nebulae whirled and then coalesced while spinning into the spheres, solar systems, and galaxies of our cosmos. TT disputes this as being far from believable.

Also, Bang and Nebular can't explain the spin seen in the macro-world of galaxies, solar systems, stars, suns, planets, moons, nor down to the quantum micro-world of atoms.

Whereas, TT shows black holes as the machine-like power engines responsible for the spinning features applied to the our cosmos in galaxies, solar systems, and spheres.

Comparing Big Bang to Trillion Theory (TT).

TT takes longer to explain than Big Bang or Nebular. TT maintains that our universe is complex and tricky, so it takes more than Big Bang's 13.7 seconds to explain.

First, here's Big Bang's explanation of our cosmos: 'Hot matter supposedly exploded from a central source across cold space, where the hot matter formed into many spheres.'

But, that leaves a zillion unanswered questions. When Big Bang failed to answer, Nebular was recruited for support. 'After the explosion, hot interstellar gases spun in space like a nebula, and as the matter spun it coalesced into the hardness of planets and moons (smaller and cooler) and into hotter larger spheres called stars. Since stars were larger, they supposed didn't cool in the coldness of space. Then, the gravity of the nebula brought the smaller spheres into orbit surrounding their suns to form solar systems and galaxies.'

Once again, zillions of unanswered questions result from the inadequacies of Bang and Nebular. Basically, we've been provided with 2 no-brainer theories. They sufficed over the past 65 years, but not today, as too many questions remain unanswered? TT flatly says "no" to Big Bang and Nebular.

Author: In some chapters of this book, an Interviewer wades in taking the liberty to pose pertinent clarification questions.

Interviewer: "From a reader's perspective, should we give a tinker's damn? Does it really matter how our cosmos started or how it became so humongous? What's in it for us?"

Author: "There exists an unstoppable human inquisitiveness to explore space, the final frontier. Ever since the caveman, we've been on a long voyage of discovery. Millennia passed between fire, spear, wheel, and printing press. But thereafter, pivotal inventions accelerated us through the ages: 1736 steam engine; 1862 plastic; 1886 car, 1890 telephone; 1903 airplane; 1928 TV; 1950 computer; and 1971 microprocessor.

What will man's next great invention be which benefits all of humanity? As a civilization, we face dramatic home-planet problems. Will man be able to find world-saving solutions?

TT bets that man's next super-inventions will come from a greater understanding of our cosmos. Will man be able to someday create new types of elements for super batteries to defeat global warming? Or use new elements to build flying saucers affording speedy space travel?"

Interviewer: "Why is your Trillion Theory important?"

Author: "Trillion Theory states: *'The knowledge of how matter was made into the atoms (elements) comprising the planets, moons and suns of our cosmos is an absolute integral part leading to the most prolific discoveries which a species such as man can ever hope to achieve.'*

Small changes can give rise to large consequences and incremental changes can turn our understanding on its head. Therefore, we need to undertake a new introspection of our cosmos, performing forensics, determined to find the truth."

Interviewer: "What can Trillion Theory explain in short form?"

Author: "While I would prefer to deliver my entire new theory in a lump sum, it takes substantially more pages than that. From chapter to chapter, TT endeavors to connect the dots, replacing old beliefs with new TT ideas. For you, it'd be helpful to keep an open mind till you've read the entire TT."

Interviewer: "What are some of these 'out-there' ideas?"

Author: "Here is a short list of new Trillion Theory ideas:

• *Trillion Years* is the age of our cosmos. Much older than the present-day estimates of 13.7 billion years.

• *Recycling* applies to all of our cosmos. TT maintains that solar systems, along with their moons/planets/sun recycled many times during the trillion years of cosmic history.

• *Black Hole prominence.* Trillion Theory claims that there is a black hole at the core of every sphere in our cosmos. These black holes are the building machines of our cosmos.

• *Light the main material.* Trillion Theory claims 'light' as the main material used by black holes to build the spheres, solar systems, and galaxies of our cosmos. We have yet to uncover all the marvelous secrets of light, and TT claims that light can be spun into matter. Up to this point in history, no human has ever 'played creator' by taking light past its bending stage and forcing it to spin and coil into atoms of matter. Such a scientific discovery would be astronomical; plus proof of Trillion Theory.

• *Grew.* Our cosmos grew from small to its present gigantic size via recycling and reproduction processes.

• *Reproductive replication?* TT uses these terms to describe how our cosmos grew from small to gigantic by deploying a recycling process. This indicates an 'alive' type of component to all cosmic moons, planets, and stars."

Interviewer: "I'm having a tumultuous time accepting a black hole at the center of planet Earth, and our sun, and inside all the planets/moons of our solar system. Where's your proof?"

Author: "No proof thus far. But, a theory placing a black hole at the core of every sphere, and a supermassive black hole at the core of every galaxy, makes a whole lot of things in our cosmos align. It makes sense that some power play between battling black holes is the actual reason for the formations seen in solar systems and galaxies. TT first placed black holes at the core of spheres in 1998. Today, TT is more convinced."

Interviewer: "Are there noted astronomers or astrophysicists who support your view of a black hole inside every sphere?"

Author: "Although my Trillion Theory began back in 1998, it is only in 2014/15 that it was published and only in 2016 that the word began to spread. Many astronomers show signs of a movement in a Trillion Theory type of direction.

For instance, on the internet go to a documentary video called *Secrets of a Black Hole by HD Universe Channel.* This video was released January 18 of 2015. The things which it says about black holes fit perfectly into my Trillion Theory. In 2015, this HD Universe Channel video had this to say: 'Black holes are everywhere in the cosmos, millions of them. A supermassive black hole seems to be at the center of every galaxy. Also, it appears they can do more than just control an entire galaxy; they also have the ability to build a galaxy.'

Now, scientists are recalibrating their respect for black holes – seeing black holes as constructive builders, not just destructive monsters. Penn State astrophysicist Yuexing Li had this to say, "Now we see black holes were essential in creating the universe's modern structure.

In 2003, a sky survey showed giant black holes centering most galaxies. Since then, researchers have tried to figure out the origin of these primordial supermassive black holes.

TT takes that a step further by declaring that there exist two types of cosmic black holes: Supermassive black holes at galaxy hubs are galaxy controllers; while lesser-sized black holes exist at the core of every star, planet and moon."

Interviewer: "OK, so where do these black holes come from?"

Author: "That's difficult to say at this stage. The same could be asked of the supposed Big Bang: who or what supplied the matter for Big Bang's supposed explosion? However, once we can solve more of the mysteries of our physical cosmos, we'll be better positioned to tackle such a query."

Interviewer: "My understanding is that the amount of cosmic matter was limited, set and fixed at origin. Therefore, black holes couldn't access new matter to build more spheres."

Author: "That's an incorrect belief. TT shows that our cosmos has a limitless supply of energy material provided to it from the outer ocean of static light, and this energy is continually pulled into action by black holes and utilized to build and increase the sphere population of our cosmos.

This is but one of the new and interesting ideas making Trillion Theory into a very POWERFUL NEW VOICE explaining how our cosmos began over a trillion years ago."

Interviewer: "Where does pure science fit in?"

Author: "In a coming chapter, you will be introduced to the incredible engine inside a black hole which creates gravity. A black hole has such intricate working parts that it must have been purposefully designed by a mind or minds with some absolute incredible super knowledge of science."

Interviewer: "You say our cosmic creator was a scientist?"

Author: "I prefer to say, used incredible science, rather than created out of nothing. I don't tackle the idea of a creator or artisan. It will be quite a time after we fully understand our physical universe that we will be better equipped to take on such questions as 'who' or 'what' used scientific genius.

I firmly believe that tremendous pure meticulous science was behind the construction of our cosmos. It was Stephen Hawking with his books *A Brief History of Time* and *The Grand Design* who brought more of a scientific approach. But, a renowned Hawking was also fooled by Big Bang.

TT unveils some of the scientific techniques which were involved behind the scenes in the building of our cosmos."

Interviewer: "Astronomers always state that light can never ever escape from a black hole. Your TT says otherwise."

Author: "I hate to say it, but astronomers are still stuck with Big Bang thinking. However, I see signs of them questioning those beliefs. When astronomers first discovered black holes a few decades ago, they were incorrect to presume that a black hole was a death trap for light. They wrongly stated, 'light can never escape from the clutches of a black hole.'

For, while a black hole does pull in and entrap light for anywhere up to billions of years, that is not the end cycle for the light. Our 10 billion year old sun, for example, has a black hole at its center which billions of years ago consumed light and entrapped light as matter around itself. But now, aging in its cycle, the sun is in the process of losing control of the tons of light which it consumed, and that matter is unraveling back to light escaping from the black hole sun.

Astronomers see a Supernova of a star finally ending as a dense singularity. They presume that the exploding and the imploding star created the black hole singularity. Whereas, TT professes that the black hole was originally the builder of the star around itself in the first place. The explosion of that star's contents, after billions of years of entrapment, exposes the naked black hole which was originally the star's builder."

Interviewer: "Black holes portals lead to other universes?"

Author: "Pure fantasy. However, I believe one day it will be possible to eclipse the speed of light, leave here - zap there."

Interviewer: "Do you think our universe was purposely built or just a big accidental occurrence?"

Author: "My hat goes into the ring with a scientific purpose. Our cosmos wasn't just a happenchance. But, on the other hand, we have been too godish. TT professes that incredible clever scientific mind(s) designed our cosmos and it wasn't a created-out-of-nothing event. Rather, our cosmos is a result of the deployment of specific materials and specific engines involved in the building process of all cosmic spheres.

Astronomers see that suns are always the central largest sphere of a solar system. Happenchance can't explain this same configuration throughout our cosmos. This means that something else is afoot, much beyond happenchance.

Therein, on a clear night, when I gaze up at the evening stars, the scene amazes me each time. The spheroids look so pristine, so clear cut with vast amounts of space in-between. Then, my eyes attempt to see past our moon's surface, deep inside to where a black hole is busy working in its inner core.

Next chapter: In furthering our understanding of black holes, we gain a far greater understanding of our cosmos."

CHAPTER 2
WHY BLACK HOLES EXIST?

Weird, mysterious. They are the least understood objects in all the cosmos. Quite simply, what we don't yet know about black holes far outweighs what we do. Prior to my Trillion Theory, no one had been able to explain their exact role in our cosmos. Prior to now, black holes were always thought of as exotic bizarre cosmic monsters. But today, astronomers see black holes as entities which bring order and organization to our cosmos. These astronomers are slowly converging with Trillion Theory.

The main reason black holes are misunderstood is that they are beguiling in their different looks. Astronomers often see naked black holes spinning light into their bowels; or, as bright neutron stars surviving the collapse of a star; or, as gigantic supermassive black holes at the hub of galaxies.

What astronomers never get to see are all the other undetectable black holes cloaked inside of and at the core of moons, planets, and stars. Trillion Theory says that a cloaked black holes resides inside of each and every sphere in our cosmos, regardless if that sphere is a moon, planet, or sun.

Yet, proof of Trillion Theory is still distant.

Of all various the hypotheses proposed by others, it is only Trillion Theory which states that light traveling close to a naked black hole is bent, curved, and then finally spun into dense matter inside of the naked black hole. As the hole devours more and more light to spin into matter, a sphere is built around the black hole. The mechanics of this process are: naked hungry black holes possess an ultra-fast spin resulting in a powerful gravitational field around themselves. This spin warps space immediately surrounding them such that anything traveling into that space, such as a beam of light, is slowed, bent, curved and then swallowed up by the black hole. TT says that swallowed light is spun into atoms of matter which are then pent up and stored inside of the body of the sphere which the black hole builds around itself. As the black hole eats too quickly and fills, it changes its duties from eating to control of its contents. Example: Planet Earth, with a black hole at its core, is presently in a long control phase. While Earth can attract new light, it is too full with no more available room to spin that light into more new matter.

Thus, TT dramatically changes how view the origin of the moons, planets, stars, and galaxies of our cosmos. TT states: *'Black holes take light and spin it into matter, holding it tight for billions of years. Therein, black holes are the spinners of light into matter inside of and around themselves to build the very spheres which exist in all the galaxies comprising our cosmos. Black holes are the builders of our cosmos.'*

Therein, TT helps unravel some of the complex mysteries of our universe by focusing on black holes and their major role in the origin and growth of our cosmos.

The history of black hole discovery.

To put it quite bluntly, tons of scientific work-time has been spent studying and theorizing about black holes, yet they are still barely understood to this day.

The concept of a body with such powerful gravity that it could swallow up light was proposed a hundred years ago. Then in 1964, Ann Ewing wrote an article called *Black Holes in Space*. However, the credit for coining the phrase 'black holes' goes to John Wheeler in 1967. Thereafter, that name stuck and black holes became mainstream research subjects.

Even then, black holes were still regarded as theoretical curiosities. However, with the discoveries of neutron stars, pulsars, and quasars - black holes gained traction.

Neutron Stars: Small but mighty, possessing rapid spin. They are the remaining cores of massive stars gone Supernova, The surviving black hole retains attracts huge amounts of light in a rush back into the hole, thereby showing a bright glow. A pulsar is a Neutron Star seen blinking on and off due to the pulsating affect caused by the stars rotation.

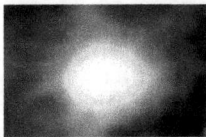

Quasars: A Quasar is the supermasssive black hole centering a galaxy, plus two twin jets pluming from the poles, plus the array of bright surrounding stars outshines a galaxy.

Naked phase of black holes is extremely hard to see.
In 1972 an astronomer pointed his telescope towards a dark spot in space detecting radio emissions. Finally, months later the dark spot eclipsed an adjacent star partially blocking out its light. An actual naked black hole had been discovered.
Then, over the years more black holes were found.
Some black holes were found to be the survivors after the collapse of massive stars which ran out of fuel near the end of their life cycle, then went Supernova, and then finally collapsed down into themselves. Plenty was done to classify the sizes of various black holes. These black holes sizes were dependent upon the star mass from which they had come.

The notion that black holes sucked in everything was correct. The area directly around the black hole was named a black hole's Event Horizon, which was the boundary that the black hole fashioned around itself in. Anything passing into this event horizon was pulled inward into the black hole. It seemed that black holes actually had the power to deform the space directly surrounding itself. Not even fast paced light racing through space could escape the clutches of a black hole. The swallowed light appeared to disappear into lockup forever. Plus, no information about that trapped light was ever again received by the outside world. In that black hole's Event Horizon, no paths led away.

Shape-wise, this gravity boundary surrounding the black hole was found to be oblate with the boundary extending out past the equator greater than from the flattened poles. We see oblate with Saturn's equatorial rings; and our solar system's shape; and the pancake shape of spiral galaxies.

Further out from the black hole, extending well out past

its Event Horizon, is a huge oblate shaped area called the Ergosphere. In this area, it is impossible for things to stand still. In TT, the powerful black hole inside our sun's core extends the drag of its Ergosphere to its planets and moons which must orbit in the direction in which their sun spins.

Question: TT, if nothing can escape the Event Horizon which is close to the black hole inside of our sun, how is it possible for light to escape and be emitted from our sun?

Trillion Theory answer: Billions of years ago, the black hole at the center of our sun filled up as it completed its eating of light. But, TT disagrees with those who say that light never escapes a black hole. TT shows how light which a black hole entraps is always destined to escape no matter how long it takes. Our sun is a prime example of an x-large black hole which won the battle to spin light into matter, but over-ate and spun loose atoms. Today, our sun is losing control of its already loosened atoms which escape as light, free to travel space. For, no sphere (moon, planet, or star) can last forever. Spheres all age-out as their matter unravels back to energy.

Nothing can escape a naked black hole, True or False?

To astronomers, a black hole devouring light seemed like a forever thing. Nothing could ever escape a black hole once it was swallowed up. It was as if a vacuum cleaner had sucked an object up forever. But in 1974, Stephen Hawking made his most famous discovery: black holes emitted radiation, meaning there was indeed something capable of escaping a black hole's clutches. And, furthermore Trillion Theory predicts that light always eventually escapes the confines of a black hole, most often taking billion of years.

But, nothing about black holes has ever been easy or quick for astronomers (astrophysicists). Each time they seem to have black holes figured out, a new fire-eating dragon rears its ugly head. Even physicist Stephen Hawking recently declared that black holes exist differently than first thought. He re-advocated that while light can't escape from a black hole, light is sort of stuck or stored in a holding pattern.

This author likes the idea that Hawking is beginning to think more like TT where light is spun into a black hole, and is stuck inside of the hole for billions of years inside of a planet, moon, or star, ready to one day eventually escape.

One certainty, astronomers/astrophysicists are looking for the next mind-bending property of black holes. Trillion Theory supplies many 'next ideas' which contradict some of the past accepted views pertaining to black holes.

Trillion Theory (TT) takes new findings a step further.

TT says we need to do a makeover and get over the preconceptions brought by Big Bang. TT says that it's high time to see black holes as cosmic builders and organizers.

TT proclaims that all the island galaxies, solar systems and spheres of our cosmos were organized by the machine-like workers known to us as black holes. These black holes built our much-ordered cosmic system.

Let's ask some pertinent questions about our cosmos. To do this, start from scratch, with open minds, and no Big Bang theory to direct our thinking. Here goes:

- Why is our cosmos so gigantic? Why not small?
- Why island solar systems and island galaxies?
- What controls formation of solar systems and galaxies?
- Why is linear formation shape seen all across the cosmos?
- Are there rules or laws seen across the cosmos?
- Why is space so vast between these galaxy islands?
- Why is space on its own so absolutely nothing?
- Why are planets cold, while stars are so hot?
- Why do planets, moons, sun spin on their axis?
- Why do galaxies spin around a supermassive black hole?
- Why is spin seen in everything, atoms to galaxies?
- Why does this spin seem to go on forever?
- What created the pull of gravity?
- Why are there black holes in our cosmos?
- Why do black holes spin at thousands of rpm, or at all?
- Why are black holes now known to be in the billions?
- Why supermassive black holes at galaxy centers?
- Why crowd of stars around a supermassive black hole?
- Why a black hole survivor after a Supernova?
- In a dark cosmos, why is there light?
- Which came first in our cosmos, light or a star?
- Why does an atomic explosion send out light?
- Does time exist across our entire universe?

TT puts forth answers, offering a new opportunity to think outside-the-box and re-think our entire cosmos. TT advice is to think strategically when examining the role of black holes.

New points of views presented by Trillion Theory:

- Trillion Theory professes that our universe is far older than 13.7 billion years old. Rather a trillion years old.
- Trillion Theory claims our universe GREW to its enormous size of 73 quintillion stars – with no Big Bang at the origin.
- TT denounces happenchance. TT claims that top scientific skills were deployed in the specific design and construction of black hole 'machines' becoming cosmic builders.
- TT places a black hole (sizes vary) at the center of every moon, planet, star, and galaxy in our cosmos.
- TT states that spheres and solar systems comprising the galaxies of our universe recycle roughly every 15 billion years. Each cycle ups the count of stars and solar systems.
- TT claims existence of an endless energy supply available to infinitely grow our cosmos. The number of stars, solar systems, and galaxies will increase exponentially forever.
- TT states that a very small galaxy can totally recycle like a solar system. But, large galaxies see a supermassive black hole at their hub for hundreds of billions of years.
- TT emphasizes the incredible properties of light as the absolute energy units deployed in the recycling process.
- TT shows that black holes self-replicate. This replication is different than anything ever seen. Yet, black holes possess qualities of growing entities, seemingly like robots. Black holes are pre-programmed with instructions enabling self-replication after a Supernova. This replication perpetually grows the number of black holes throughout the cosmos. This feature given black holes by TT may seem tantamount to science fiction. Yet, TT likes to show that our cosmos is all about the very tops in science. Black holes fit that bill.

What do some modern astronomers have to say?

A group of top astronomers do work at the Mt. Laguna Observatory in California. Their main jobs are to study black holes of all types and sizes. Their modernistic thoughts about black holes are most interesting. Here is a list of what they had previously thought about black holes:

- black holes were thought to be a rare cosmic feature.
- black holes were difficult finds in the dark of space.
- black holes fostered terrifying violent drama.
- black holes brought only chaos and destruction.
- black holes were the most destructive cosmic objects.
- black holes could rip apart a star.
- black holes could make a star go Supernova.
- black holes were a one way street for captured light.
- black holes could swallow all nearby light and matter.
- black holes could spin furiously.
- black holes all spun in only one direction.

Now, these top astronomers at Mt. Laguna Observatory are unlocking more of the secrets of black holes:

- black holes are not a rare cosmic feature.
- black holes are becoming easier to find in dark space.
- there are billions of black holes in our cosmos.
- the fact there are so many black holes is extraordinary.
- black holes are the most mysterious cosmic objects.
- so far, black holes defy understanding.
- black holes are not just chaos and destruction.
- black holes are not an eternal one way street for light.
- black holes can spin at millions of miles per hour.
- black holes can spin in either direction.
- black holes stage the most violent battles in the cosmos.

- but, black holes also have another gentler side.
- supermassive black holes are at the center of galaxies.
- these supermassive act as anchors for a galaxy.
- supermassives control the organization of a galaxy.
- black holes have a purpose of being cosmic organizers.
- black holes seem to shape our cosmos.

Author: In some chapters of this book, an Interviewer wades in taking the liberty to pose pertinent clarification questions which have been asked of the author at science expos.

Interviewer: "You say that scientists haven't figured out why black holes exist or how they fully operate?"

Author: "Exactly. You see, the monumental discovery of black holes was just a first step in uncovering their secrets: such as, why they devour light? Astronomers incorrectly thought of black holes as simply rogue-like invaders of space, looking to devour a star, for no good apparent reason. New TT theory gives black holes a whole new importance, making them prevalent throughout our cosmos and instrumental centers of birth and rebirth of every cosmic sphere.

TT says, 'Black Holes are the power engines which built the moons, planets, and stars of our cosmos by the spinning of matter around their cores.' As engines, they exist mainly hidden away at the central core of the sphere which they spun around themselves. However, a black hole can go back to naked by shedding all of its bodily matter in a number of ways: as a sun emitting all of its light rays; as a Supernova; or as a sphere destroyed during the Obliteration of a solar system. Note: Black holes also exist in a supermassive size at the hubs of colossal spiral galaxies. See the Galaxies chapter.

Just like $E=mc^2$, where both energy and matter cannot be destroyed but can only be interchanged, black holes are the master-survivor recyclers of our cosmos. They can never be destroyed even by the power of a Supernova. Try to destroy a black hole and it simply splits replicating to two new black holes. Each then builds a new sphere of matter around itself.

Over the last trillion years, black holes have been *sphere factories* hard at work building and recycling the spheroids of our cosmos. A fundamental component of Trillion Theory is how our universe originated. Today's enormous cosmos is a product of black holes being recycling sphere factories."

Interviewer: "Where did the original static ocean of light and that initial black hole come from?"

Author: "That is the huge question, to be sure. To know for certain one would need to go back to the Artisan creator of this universe (if such an entity exists), knock on the door, and make that query. We have been provided with a cosmos which functions well. As to where the materials came from, I've always loved a mystery. Perhaps whoever or whatever has been cleverly egging us on, tossing us helpful clues such as rainbows, lightning, nuclear explosions, and supernovae.

Let's just say that any artisan inventor of our cosmos is staying in the background for now and allowing the cosmos do its thing according to the 'instructional imprinted laws' laid into the building material light, and engine black holes.

It is through the following and mapping of the life cycles of black holes that science can prove Trillion Theory. Black holes are the engines of the building and the snail-paced recycling of the spheres in the solar systems in our cosmos."

CHAPTER 3
OUTSIDE (INSIDE)
A BLACK HOLE

Since black holes are absolutely vital to Trillion Theory, it's imperative to closely examine the specific makeup of black holes which allows them to perform their miraculous tasks as builders and organizers of our cosmos.

As previously stated, what we don't know about black holes far outweighs what we do know. No one has yet ever seen into a black hole. Obviously, this is a mysterious and possibly an extremely dangerous spot. We can only fantasize about entering into a naked black hole, and that fantasy might have horrific consequences when encountering such ferocious spin. It takes something as indestructible as 'light' to survive such a dangerous passage.

This chapter first describes the outside area around a black hole. Then, we venture inside a black hole to see what has never been seen before, namely the *engine structures* causing the events inside and around a black hole.

Observing black holes.

Today, astronomers detail black holes as commonplace in our cosmos. They see naked black holes eating light; as well they see a black hole remnant after a Supernova explosion; and they see supermassive black holes at galaxy hubs. But, astronomers have yet to discover a black hole at the core of any intact planet, moon, or star. Yet, TT says "look harder."

TT advocates that there is a cloaked black hole, hidden away, at the core of every sphere including planet Earth.

The one residing at the core of our planet is hidden away under far too many tons of matter to be detected. The only time our black hole will uncloak is when our planet dies and sheds all of its matter, thereby exposing its black hole.

Therein, TT contends that cloaked black holes residing at the core of spheres far outnumber the visible black holes of our cosmos. To approximate the number of cloaked black holes throughout the entirety of our cosmos we calculate as follows: 73 quintillion total stars X approximately 50 planets and moons per each star with a solar system = a staggering 3,650 quintillion cloaked black holes are hiding away and masquerading inside of a the spheroids of our cosmos.

But, first we examine a naked black hole before it became cloaked and hidden at the core of a spheroid.

An astronomer's outside view of a naked black hole.

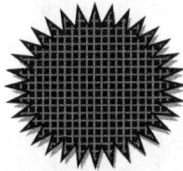

The naked black hole appears inky black, spinning at great speed on its axis. Its rough surface provides it with better grip for pulling nearby light inwards.

Next, TT examines the area surrounding a black hole.

Naked black hole is surrounded by an Event Horizon, then further out an Ergosphere, and then the vastness of space.

The inner area is of course the black hole; absolutely dark and spinning at tremendous velocity. When empty, TT calls it a naked black hole which is starting to feed by pulling light into itself. A naked hole has great spin speed around its axis creating a centrifugal force out to the surrounding area.

Directly surrounding the naked black hole is the Event Horizon where a black hole displays its strongest influence being able to curve space inwards to pull light into the hole.

Further out is the Ergosphere, where the extended gravity pull of the black hole can hold other spheres in orbit.

Comparing the 'Event Horizon' and the 'Ergosphere' of a black hole as Trillion Theory sees them:

The Event Horizon is a small powerhouse area starting at the center of a black hole and extending outwards directly around the black hole. Events occurring within the Event Horizon are the most significant of any cosmic zone. For, TT alleges that the Event Horizon is where light is attracted, bent, and then bent to the extreme being spun into matter. Event Horizon is where the creation of matter occurs.

Trillion Theory reveals that Event Horizon is where light is taken past curving and bending, to the spinning point. This differentiation (made only by TT) has huge repercussions as physicists studying light have never yet been able to spin light into matter. TT suggests that light can be spun into matter and shows how black holes are capable of the feat.

This light spun into matter is locked away inside of atoms inside and around the black hole for what seems like forever. However, escape always comes for this spun light, although the process of the loosening and unraveling of atoms, and the eventual escape of light, normally takes billions of years.

Now, getting back to the eating habits of the naked black hole, if there is a large supply of available light the naked black hole will quickly fill its belly full of freshly spun atoms of matter. Then, the now less naked black hole will add even more atoms to begin building a body. Once full all the way to its periphery surface, the black hole will be cloaked inside of voluminous matter. A planet or moon has formed.

In the aftermath of this spinning of light into matter, for the following billions of years, the black hole's Event Horizon can still attract light to the sphere, but it can no longer spin light into matter. Example: Planet Earth's black hole is full.

This Event Horizon is a strong gravity area where objects are pulled down to the surface of the spheroid.

Now, the black hole turned spheroid focuses its attention on controlling its spun matter from escaping its surface. This hold is so strong that it extends far out past its surface to a large space area known as the Ergosphere. Here, gravity is strong enough to hold other spheres in orbit.

The outer limit of this Ergosphere depends directly upon the size of the black hole and the mass of the sphere which the black hole erected around itself. The Ergosphere of our sun projects outwards billions of miles allowing it to hold many planets and moons in orbit.

In Ergosphere all objects must be in continual motion as caused by the black hole's powerful spin. As it drags the Ergosphere around itself, all the spheres within Ergosphere revolve in orbits which travel in a similar direction as the black hole's axial spin. Our sun has a counterclockwise axial pivot, so all the spheroids caught in its Ergosphere orbit around our sun in a counterclockwise direction.

We examine the **Ergosphere belonging to the sun of our solar system**. TT states 'that there is a cloaked black hole residing at our sun's core.' This naked black hole was naked empty before it began to build our sun. Since that black hole was the largest in this region, it easily won the battle for light against lesser black holes. It gobbled, spinning atoms more loosely than any of its lesser more patient competitors. As this black hole spun more light into tons of matter, the black hole became hidden and cloaked inside of our sun.

Then over billions of years, the black hole over-filled as a result of its gluttonous eating. It became sluggish with lower spin speed. Overtime, it began to lose control over its atoms which began to unspin back to light. Eventually this unravel process became so intense that the sphere turned into a blazing star with light escaping the surface at a frantic rate.

However, the black hole at our sun's core still possesses enough of a powerful gravity force out into its Ergosphere to hold all the spheres of our solar system in continuous orbit.

Next, we examine the **Ergosphere of Planet Saturn**, and the shape of its majestic rings. TT maintains the existence of a cloaked black hole residing at the core of Saturn.

For Saturn, this black hole began naked, but as it spun light into tons of matter around itself, the black hole became hidden inside. Not as large as our sun, yet not as small as many of the lesser sized black holes of the region. This black hole spun light into atoms of medium tightness, producing more of a gaseous, rather than a hard surfaced planet.

Saturn's black hole is able to hold 62 moons in orbit in its Ergosphere. The rings of Saturn illustrate a feature showing how a black hole's gravity projects strongest at its equator.

Evolution-wise, Saturn has become a gas giant, meaning that its black hole has lost some of its hold on its matter. So Saturn is less solid than Earth, but not as gaseous as a star. It may someday evolve into a star, but more likely it will have its matter destroyed when our sun goes Supernova meaning that the black hole at the core of Saturn would survive and return to being a naked black hole eater and spinner of light.

Lastly, we look at the **Ergosphere within a spiral galaxy**. The supermassive black hole at the central hub of a galaxy has evolved, as it displays some new and different talents. A main feature is that a supermassive no longer finds the need to have to grow a large body of matter directly around itself. Rather, it relies on its Ergosphere to accomplish its tasks. The Ergosphere of the supermassive can extend billions of miles out into its spiraling arms. Note: Supermassive black hole features are discussed in greater detail in later chapters.

We go inside, into the bowels of a black hole.
What might be the inside structure of a black hole?

Black Hole Structure

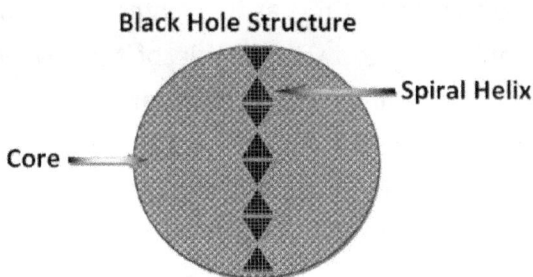

**A Black Hole in its naked phase. It has a spinning 'Spiral Helix'
running through its middle and a 'Core of Compartments'
(each spinning) as main parts of its roundish design.**

TT speculates as to the internals of a naked black hole. The interior axis of a naked empty black hole has the fastest cosmic spin. TT envisions an internal structure allowing this to occur. Such structure must account for a naked black hole's talents, namely: the ability to use its fast spin to inwardly curve an area of space around the naked black hole; to attract light into that area; to take attracted light past its bending rainbow property and spin light into matter.

The reason a black hole can achieve such a high spin rate is the structure of its axis. Compare to a spinning toy top.

The pump toy top uses a pump handle to send a fast spin to its body. The harder and faster the handle is pumped, the faster the body spins. As the handle depresses down, the body is imparted a spin, as the handle raises the body rests in spin.

A main key to black holes is their pumping helix axis. It works similar to the twisted rod at the center of a small toy called the pumping toy top. The handle of the toy pumps down and then up, moving a spiral rod in the center of the toy, thereby creating spin. The spiral rod axis then imparts motion to the body making the toy top spin on the floor.

In a similar manner, the helix axis of a black hole uses its elastic property to lengthen and shorten the helix thereby causing the core to rotate. To perpetuity, a shortened helix has a need to lengthen and a lengthened helix must shorten.

Let's look at a naked black hole, 100% empty, spinning at top speed, ready to spin light to matter. Its goal is to build up matter. Often, other black holes are in direct competition.

The naked hole begins spinning at a fantastic unabated speed. Its core body, with a myriad of compartments, spins around its rod-like spiral helix. Phenomenal elasticity of the spiral helix pumps up and down, imparting a spin to the spiral rod. This spiral helix then imparts this motion to the body making the entire black hole spin on its axis. Linear spindle motion torques into circular spin of the entire body. The naked black hole is now ready to begin feeding.

An important part of the structure of a black hole is its tentacle filaments. Every compartment has one to aide in the pulling of the light inwards into the bowels of the black hole.

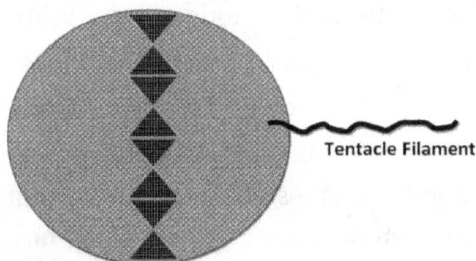

A *tentacle filament*, as one example of millions, extends from a compartment of the interior body of the black hole. These myriads of filaments utilize their grasp to aide in the securing and pulling of light inwards into the core of the black hole.

The fast spinning motion of a black hole projects a strong pull into the *Event Horizon* surrounding the black hole. Any light contacted within that Event Horizon can be pulled inwards, bent, and finally spun into matter inside of the black hole.

Inside of a naked black hole, processes are occurring.
Light, pulled inwards from the surrounding event horizon, is taken deep into the core of the naked black hole. The first attracted light will make it all the way to the spiral helix and the very center of the black hole. In TT, the laws pertaining to black holes show that the spiral helix will fill first, able to grow its length. Note: This is evolution and survival of the fittest. Black holes are programmed to become as large and powerful as possible in direct competition against all other black holes within their vicinity.

Next, the black hole will spin light into matter to fill the compartments of its main interior body. With this fill up, each compartment utilizes its elastic feature to stretch and become larger. At maximum stretch, the compartments can subdivide, doubling compartments. With filling, stretching, and subdividing techniques, the black hole can grow larger.

What happens when a once-naked black hole has filled up its entire interior with matter spun from light?
TT shows that a black hole will continue to spin more light into matter, even after it has totally filled its interior. Over time, the naked black hole becomes less visible as it becomes fully dressed and cloaked as it spins tons of light into a body of matter. Eventually, the black hole builds itself into a moon or planet. The size of the resultant sphere is most dependent upon the size of the black hole doing the building. As the black hole becomes totally full with matter, it can still attract light to its Event Horizon, but the full black hole will no longer spin that light into matter. Passage is blocked by all of the matter built up from core to surface.

While spin rate of a naked black hole is astronomically

fast, TT maintains that the rate at which a black hole spins will slow down drastically as the black hole builds more and more of a body of matter around itself. This massive body carries such weight and mass that the black hole residing inside is forced to pivot at a much reduced rate on its axis.

The cloaked full black hole now uses its slowed spin and the weight of its bodily matter to show-off its gravitational pull. The cloaked full black hole now enters up to billions of years where its prime concern is utilization of its gravity to control its atoms of spun matter. The power of this gravity gets extended past its surface out to spheres held in orbit.

What happens if more than one naked black hole is at the same time filling up in the same locale?

Imagine many naked black holes located where a new solar system is about to be built. They are the survivors of the Supernova of their old sun and an Obliteration of their old solar system. With Obliteration, planets/moons melted, going Supernova as well. These naked black holes are now all that is left of the spheres of that old solar system.

All of these newly exposed (now naked) black holes will engage in battle for available light. Survival of the fittest (size, speed, and location) will determine the final results. The *Laws of Naked Black Holes* and *the Laws of Spheres* will determine the organized make-up of the new solar system.

Interviewer: "I find it very intriguing how you have come up with the mechanisms inside a black hole."

Author: "Thanks, but although TT put forth ideas about black hole mechanisms, it may be eons before astronomers find a means to examine the intricate parts of a black hole. Those parts will likely be something far beyond our imagination."

Interviewer: "You speak of a wondrous mechanism allowing a single black hole to split into two new black holes?"

Author: "Conjecturing black hole replication is difficult. Black holes are the most powerful cosmic engines. Whereas, the most powerful events are Supernova (where a star explodes) followed by Obliteration (explosion of a solar system).

Prior to a Supernova, the black hole in the central core of a star has been discernibly witnessing the unraveling of its atoms for several billions of years. Once this unravel begins, a black hole can never commandeer it to stop. A star's body mushrooms, going Supernova exploding all of its longtime pent-up atoms back to straight line light.

In time-lapse, as the explosion occurs, the emptying black hole regains its terrific spin speed. In conjunction with the explosion, a reaction occurs. For, TT affirms that a black hole cannot be destroyed, but a black hole can be split from a Supernova's force. An instantaneous backlash occurs. (For every action there is an equal and opposite reaction).

This powerful instant counterspin places enormous force on the black hole's spiral helix, splitting the helix in two and the entire core of the black hole as well. The two split parts of the black hole shove hard, away from each other.

Two new black holes replace the old one. Immediately the two new naked black holes begin competing to devour light. They have gone immediately back to their virgin impulses. (See the chapter on Replication)."

Provocative thought has always led the way. Science runs in behind to prove or disprove. No great idea ever first began without being thought of as outrageously ridiculous."

CHAPTER 4
HISTORY OF TRILLION
YEAR COSMOS

Declaration by TT: At a trillion years, our cosmos is much older than other 13.7 billion year estimates. Far older than the stars we see in the present sky. Using powerful telescopes, astronomers have catalogued some of the oldest stars in our universe to be in the 13 to 14 billion year range. From this viewing evidence, they estimate that the age of our universe is 13.7 billion years because that's the age of these oldest stars. They do not realize that the universe knows how to reseed itself from old dying stars to recycle over and over again. And Trillion Theory says that our universe has been doing this for a trillion years.

Age crisis in the cosmos.

In 2007, a unique star was observed to be 18 billion years old, providing a puzzling quandary for astronomers as to how a single star could be older than their supposed 13.7 billion year old universe. A similar paradox continued in 2013 with the discovery of the impossible Methuselah star, estimated at nearly 16 billion years old.

The Answer - Can't see the cosmos for the stars.
A Comparison to: Can't see the forest for the trees.

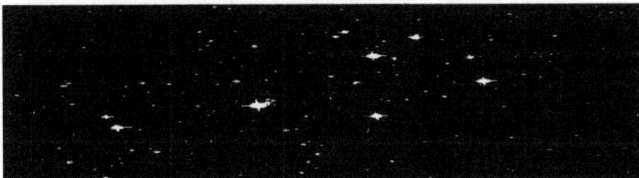

A great analogy can be drawn between the trees in a forest and the array of stars in the cosmos. In a forest, we see trees of various ages: old and young, with the oldest perhaps 200-300 years. We incorrectly conclude that forest as a ripened 300 years of age. Yet, that forest has recycled hundreds of times and is millions of years old. Proof is found in old dead trees rotting away deep below the forest's floor.

However, the ancient history of cosmos is better hidden within the recycling process of its spheres. Stars live so much longer and age so slow, that humans totally misread them. Like trees, stars also have a recycle life. Stars experience a black hole birth followed by a long life with a general upper limit of 15 billion years. Stars, planets and moons die and then perpetuate as new spheres into the next 15 billion year cycle. Star populations and also the number of solar systems inside of galaxies increases with each 15 billion year recycle.

People incorrectly miscalculate our universe's age using the oldest present-living stars as their gauge. Unfortunately, no star graveyard has left markers behind as evidence. Old dead stars so wondrously and completely recycle leaving no seeable traces (except for naked black holes ready to form the next solar system in that locale). Remnants of the ancient stars and solar systems are near impossible finds. Each old star cycle is lost into history, while each new cycle appears unique providing the illusion of being the one solitary cycle.

Trillion Theory says that we are presently living in the 67th of the 15 billion year cycles of our universe. But, for us there is the illusion that we have been the one and only cycle.

Star population of our cosmos.

In this 67 recycle of the stars, to approximate the number of stars, multiply two hundred billion galaxies (approximate) by three hundred and seventy million stars per galaxy = over 73,000,000,000,000,000,000 (quintillion) stars. Trillion Theory follows this tally from present-day astronomer's estimates.

Size our Universe.

Our universe is trillions of miles across. An easier way to explain that distance is in light years, which is the distance which light travels in a single year. It takes eight minutes for light to get from our sun to Earth, so light travels a long way in our cosmos in one light year. Yet, our cosmos is far too big for light to move very far across its expanse in a single light year. Our cosmos has grown with each recycle to being hundreds of billions of light years across.

Correlation between population, size and age.

During universe history, stars increased in number each recycle, as did the spatial size of the universe. This took time, hundreds of billions of years. Star population, spatial size, and universe age are all intertwined and interdependent. But, our cosmos doesn't really show its true trillion year age, hiding behind the present cycle of stars.

Age of our physical universe.

While Trillion Theory calculates our cosmos at a trillion years of age since inception, that estimate is based upon growth starting from one solitary star. However, Trillion Theory concedes the possibility that initial onset of our universe may have begun with hundreds or thousands of initial black holes forming into stars. Therein, such a faster start could have taken 100 or 200 billion years off the age.

However, Trillion Theory mathematical calculations begin with just the one star and grow from there with each recycle.

The following mathematical progression shows doubling of the star population of our universe each 15 billion years. The numbers start from 1 star in cycle 1, doubling with each new cycle, and dramatically escalating at our 67th recycle.

Cycle 1 Stars 1; Cycle 2 Stars 2; Cycle 3 Stars 4; Cycle 4 (8); Cycle 5 (16); Cycle 6 (32); Cycle 7 (64); Cycle 8 (128); Cycle 9 (256); Cycle 10 (512); Cycle 11 (1,024); Cycle 12 (2,048); 13 (4,096); Cycle 14 (8,192); Cycle 15 (16,384); Cycle 16 (32,768); Cycle 17 (65,536); Cycle 18 (131,072); Cycle 19 (262,144); Cycle 20 (524,288); Cycle 21 (1,048,576); Cycle 22 (2,097,152); Cycle 23 (4,194,304); Cycle 24 (8,388,608); Cycle 25 (16,777,216); Cycle 26 (33,554,432); Cycle 27 (67,108,864); Cycle 28 (134,217,728); Cycle 29 (268,435,45); Cycle 30 (536,870,912); Cycle 31 (1,073,741,824); Cycle32 (2,147,483,643); Cycle33 (4,294,967,296); Cycle 34 (8,589,934,592); Cycle35 (17,179,869,184); Cycle 36 (34,359,738,368); Cycle 37 (68,719,476,736); Cycle 38 (137,438,953,472); Cycle 39 (274,877,906,944); Cycle 40 (549,755,813,888); Cycle 41 (1,099,511,627,776); Cycle 42(2,199,023,255,552); Cycle 43(4,398,046,511,104)

Cycle 44 – Stars 8,796,093,022,208

Cycle 45 – Stars 17,592,186,044,416

Cycle 46 – Stars 35,184,372,088,832

Cycle 47 – Stars 70,368,744,177,664

Cycle 48 – Stars 140,737,488,355,328

Cycle 49 – Stars 281,474,976,710,656

Cycle 50 – Stars 562,949,953,421,312

Cycle 51 – Stars 1,125,899,906,842,624

Cycle 52 – Stars 2,251,799,813,685,248
Cycle 53 – Stars 4,503,599,627,370,496
Cycle 54 – Stars 9,007,199,254,740,992
Cycle 55 – Stars 18,014,398,509,481,984
Cycle 56 – Stars 36,028,797,018,963,968
Cycle 57 – Stars 72,057,594,037,927,936
Cycle 58 – Stars 144,115,188,075,855,872
Cycle 59 – Stars 288,230,376,151,711,744
Cycle 60 – Stars 576,460,752,303,423,488
Cycle 61 – Stars 1,152,921,504,606,846,976
Cycle 62 Stars 2,305,843,099,213,693,952
Cycle 63 Stars 4,611,686,018,427,387,904
Cycle 64 Stars 9,223,372,036,854,775,808
Cycle 65 Stars 18,446,744,073,709,551,616
Cycle 66 Stars 36,893,488,147,419,103,232
Cycle 67 Stars 73,786,976,294,838,206,464
(Over 73 quintillion stars)

Astronomers believe there are probably an even higher number of stars; somewhere over 1 sextillion (21 zeros). This could make our cosmos even older.

Proclamation: Our cosmos, according to TT, is one trillion years old; 67 times older than the 13.7 billion year estimates. Cosmos is currently celebrating its trillionth birthday.

Variance age of stars, solar systems and galaxies in TT.

Each of the 67 cycles of our cosmos lasted approximately 15 billion years. There is no exact cut off between one cycle and the next simply because star-lives blend and overlap into the next cycle, with threads from one cycle to the next. This occurs because of the variances in life cycles of stars.

But in general, 15 billion is the normal length of the life

cycle for a star. Therein, 15 billion years is also the normal length of the normal life cycle of any solar system as its contents recycle along when the star goes Supernova.

But, an ageless galaxy follows different rules because of the supermassive black hole at its hub which lives by a different set of cosmic laws. Some galaxies are younger at 100 or 200 billion years, while some are 500-800 billion years old. Because different solar systems within a galaxy recycle on different time schedules, an entire galaxy never has to totally recycle all of its solar systems at any one particular point. So, while galaxies are timelessly old, their stars recycle intermittently each 15 billion years, just not all at once.

Implications of a recycling cosmos.

From this, draw two dramatic conclusions. Firstly, cosmos had an inception. It started small, has grown, recycled, increased in population and expanded in size with each new cycle. Secondly, cosmos will perpetuate to infinity; matter and energy will go back and forth through eons of time.

Also, the cosmos will continue to draw new allotments of static light available from the outer ocean of static light to be spun into more new matter. Because of its method of construction, this cosmos is designed and programmed to recycle, perpetuate and grow onward into eternity as it has no possible end. It cannot be destroyed. Not by any means known of, and certainly not of its own accord.

From now on, when you see a ray of light leaving our sun, you may see it differently, as light caught in a recycle which has a long ancient past stemming back a trillion years.

As a ray of light leaves our sun, it passes by an endless unnoticeable graveyard where there once existed spheres of previous cycles. In a historical sense, it is unfortunate that we can't see this old graveyard as evidence of those ancient dead spheres. Our universe's long history has been erased by recycles so complete that the remnants of the ancient past are nowhere to be found. Each old cycle is lost forever, while each new cycle gives the illusion of being unique.

The brightness of a Supernova is hitting us over the head to get our attention. Such a bright explosion is saying, 'Look here! See this exploding star; see the obliteration of a solar system; see the after effects as new naked black holes fight for light to spin as they begin birthing a new solar system.

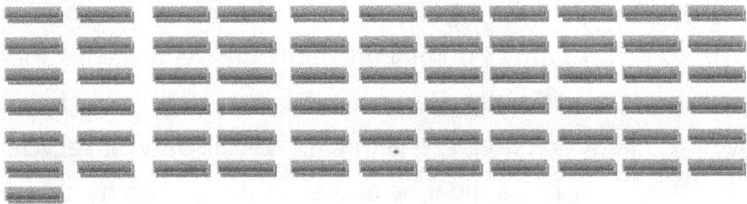

Trillion Theory ages our universe at a trillion years represented by the 67 bars (each bar equals one 15 billion year time frame)

Each 15 billion years is one cycle. (A cycle can have some overlap with a preceding or following cycle). Take our solar system as an example, we are about half the way through our present 15 billion year cycle of our sun and after our sun goes supernova, another 15 billion year solar system will grow from the surviving black holes which are at the core of every sphere in our solar system. However, neighboring solar systems might be on a totally different recycling schedule.

Therein, ties always exist and overlap between one 15 billion year cycle, its previous cycle, and the coming cycle.

Giving our cosmos a deserving name - Recycliun.

We could call our universe 'old indestructible.' A better name also describes operations. TT likes to call our universe 'Recycliun' composed of key letters taken from three key words, namely: Recycling (take Recyc) meaning that our universe recycles; from the word light (take li) indicating that the substance recycled is light; and word universe (take the un) stating that our universe is where the light is recycled. Recyc-li-un, its action and duration is Relunoty (**Re**cycling light **un**iverse **o**ne **t**rillion **y**ears).

Trillion Theory shows how Recycliun operates.

It's actually quite ingenious how our Recycliun cosmos grew to its present 73 quintillion star count via its own super recycling methodology. Recycliun has recycled many times over. It has doubled in size with each 15 billion year recycle, becoming ever more gigantic and expansive. Since its origin, Recycliun has witnessed population growth with a doubling in size from one 15 billion year cycle to the next.

Our present 15 billion year cycle of our cosmos is but the latest in a lengthy history. Past recycles were so complete that all evidence of the near agelessness has vanished. The reason being that neither matter nor energy can ever be destroyed as they simply recycle back and forth; over and over leaving nothing behind as identifiable evidence.

Where old stars go to die.

Why don't we see long term graveyards piled full of old dead stars and star dust? No, that doesn't happen, because black holes never die. TT maintains that the Supernova of a star never spells death for the black hole at the star's core.

In the aftermath of a Supernova, the now uncloaked black

hole which was at the core of the destroyed sun survives and is ready to begin its new assignment. The now naked black hole immediately sets to work by attracting new light and spinning that incoming light into new matter to form a new bodily sphere around the black hole. This re-building might take a single year or hundreds of years, dependent on how plentiful the supply of light. Once again the black hole fills up and re-cloaks itself inside of a new sphere. Therein, any evidence that the old star ever existed is lost forever.

Replication of a black hole follows a Supernova.

While the star is gone, its black hole core remains intact. The Supernova star exploded, unraveling back to light all the atoms which had comprised its former body.

There are two indestructible materials in cosmos which can never be destroyed, namely light and black holes. When Supernova occurs, the extreme action of the death of the star forces its black hole core into a splitting action from one to two. Hence, two twin naked black holes replace the old one. Now twice as many replicated naked black holes can grow into twice as many new spheres.

What if our sun went Supernova, destroying our sun and the 8 planets and the over 100 moons our solar system. All the spheres would be wiped out, but the black holes at the core of every sphere would survive and replicate to double the number of naked black holes ready to fight for light in the process of erecting a new solar system.

This naked black hole replication is a new discovery of TT and the chief reason cosmos is populated with quintillions of planets, moons, suns, solar systems, and galaxies.

Interviewer: "Please, how is our cosmos a trillion years old? Astronomers say our universe is 13.7 billion years of age."

Author: "TT contends that we presently reside in the latest 15 billion year visible cosmic cycle, There have been 67 of these intertwining 15 billion year cycles throughout cosmic history; 15 billion X 67 recycles = 1,000,050,000,000 years. So, Big Bang Theory incorrectly estimates the age of cosmos as 13.7 billion years. In stark contrast, TT's estimate is far older, set at a trillion years. TT depicts a history comprised of 66 past eras with an era 15 billion years in duration. We are presently living through the 67th cycle of 15 billion years."

Interviewer: "Why is 15 billion years so critical?"

Author: "Please don't get too hooked up on that 15 billion year number. It's an approximate life time for a star and its solar system; dependent upon many factors. Ex: A sun strives to live a long life, but the accidental entrance of a rogue black hole into a solar system quickly shortens a sun's life. The history of Recycliun has been: One cycle, two cycle, three cycle, four; marching ever onward over the past trillion years, one approximate 15 billion year cycle after another.

It's actually quite ingenious how our cosmos grew to its present 73 quintillion star count via its own super recycling methodology, doubling in size each recycle. Our present 15 billion year cycle is but the latest in a lengthy history. Unfortunately, past recycles were so complete that evidence of cosmic agelessness was perfectly recycled away.

The reason is that matter and energy recycle back and forth, leaving nothing behind as identifiable evidence. Since neither matter nor energy can ever be destroyed, they have the endless ability to recycle back and forth to eternity."

Interviewer: "Whoa, while I understand your recycle concept, I'm not accepting how you say cosmos has grown..

Author: "Recycliun methodology as to how it grew so mega, *Relunoty* (**Re**cycling **L**ight **U**niverse **O**ne **T**rillion **Y**ears).

While I state that all the stars and planets in our cosmos recycle, it's really light which recycles. When a star dies, that same exact star does not recycle. Rather, the escaped light from a star travels space, available to be recaptured and recycled by waiting black holes, to grow the next sphere."

Interviewer: "So, when light comes from our sun to planet Earth, the Earth recycles that light?"

Author: "No. Earth, at its present age, stage and phase, can only absorb or reflect light. That is a different process. For light to be recycled to grow into a new planet, the planet must be at a far earlier stage in its growth phase with the black hole at its core still devouring light and spinning that light into a sphere of matter around itself. The black hole living at the core of our planet Earth has already completed that task and is now in a holding of matter pattern."

Interviewer: "So, you say there are 67 different universes?"

Author: "No. You've been reading far too many books about alternate universes. While I have no doubt different kinds of universes exist, Trillion Years Theory only speaks about this universe we presently live in and how it is much older than we have been led to believe. We reside in the 67th of the 15 billion year cycles of the trillion year history of our cosmos. Therein, ties always exist and overlap between one 15 billion year cycle, its previous cycle, and the next cycle."

Interviewer: "Do your radical theories have proofs?"

Author: "Read Chapter 17 which deals with proof."

CHAPTER 5
THEORY OF LIGHT FIRST

Which was first, star or light? Everyone, following their eye would say that's easy, we see our sun and then we see light coming from that star. So, people answer that the star was first. But, Trillion Theory disagrees, stating that before there were stars or space in our cosmos, there was only light. Light was around before the stars ever formed. In TT, this is known as Light First.

This light was endless as a vast infinite ocean of light. So, before galaxies, solar systems, and spheres formed in our cosmos, first there was only a never-ending static ocean of light. Even the frontier of space was yet absent. This original light was frozen-like, stacked like lumber, as a non-moving static ocean of light, ready for use. Note: In 2014, physicists froze a light ray for the first time, slowing its speed to zero.

Before galaxies, solar systems, and spheres formed in our universe, first there was only a never-ending static ocean of light. Even space was yet absent. This original light was frozen-like, as a non-moving static ocean of light, ready for use.

This static non-moving ocean of light held tightly packed light strands. All the material necessary for sphere-building was waiting and contained within each strand of light in the stack; namely the light spectrum, heat, weight, and mass.

Then, the first naked black hole (the engine for building) was introduced into the static ocean of light a trillion years ago. Envision a black hole interjected into the ocean of light, a naked black hole set to consume light from the ocean of light. The commencement of the building of the very first sphere of our cosmos had begun. Immediately, that naked spinning black hole began to loosen and break light strands away from the ocean of light. The naked black hole quickly consumed the broken away light and spun those light rays into atoms of matter inside and around the black hole. Huge amounts of the strands of light were consumed from the ocean of light by the black hole. Soon, a round-bodied sphere began to build around the black hole. The first planet of our cosmos was forming. It had weight and mass which light supplied, and spun on its axis around the black hole.

Back at the ocean of light, from where huge amounts of light were taken, a vacant area was seen. This vacant space surrounded the planet, but this space now lacked light, heat, weight, and mass. This space occupied a large vacated area inside the still prominent static ocean of light. Space became the weightless empty black void which light used to occupy.

Within the bounds of the static ocean of light, the first naked black hole was interjected. Immediately, that spinning black hole loosened, broke away, and consumed light from the ocean. Those light rays were quickly spun into atoms of matter inside and around the black hole forming a sphere. That sphere had weight and mass which light had supplied. And, black space became the weightless empty area which light used to occupy.

To capture still more light, the black hole worked its way continually deeper into the ocean, consuming tons of more new light. As the black hole gnawed away at the ocean wall of light, a strand of light (broken free) tried to speed away. However, that freedom only lasted a second as the black hole captured and spun that ray of light into matter.

Eventually, the black hole ate its fill of light, forming an entire heavy sphere around itself. This sphere of formed matter spun on the axis of the black hole and continued to project a gravitational pull as the interior hole still attempted to pull more light in towards itself. But now full, the black hole could only attract light to its surface, unable to spin any more light into matter. The black hole went from its phase of spinning of light into matter, to a holding pattern for billions of years. It would attempt to maintain control of the tons of light which it had spun into uncountable atoms of matter.

The spun light would now be jailed as matter for billions of years, waiting to someday escape to freedom and able to travel through space as free running straight line light. That someday would be billions years until a time when the black hole's holding abilities would tire, and then atoms of matter would unspin and escape as free traveling light.

Indestructible light recycles back to free traveling straight line light after several billions of years of entrapment inside an atom. Light finally escapes the clutches of the black hole.

In summary, light preceded the star, and outlived it too. The catalyst naked black hole ate from the ocean of light, forming atoms of spun light into a sphere, leaving a totally vacant area called space between itself and the light ocean.

TT and Light First Theory provide a new big picture.

There is a need to really rethink our cosmos. We must now view light, matter, and space somewhat differently. In Big Bang, we were taught that matter exploded cross space. Whereas in TT, material light was first as a static ocean and naked black holes accessed this light to build the spheres of our cosmos; space is simply a left-behind empty highway.

So in TT, at the cosmic origin, there existed only light in every possible direction; a static light as complete and as voluminous as the water filling an ocean.

The amount of light needed to spin the atoms and the body of matter around a sphere is immense. Formula $E=mc^2$ calculates the huge ratio amount of light energy which goes into a relatively small amount of formed matter.

TT illustrates how our cosmos began small by forming a single sphere and then a tiny solar system a trillion years ago. They multiplied growing into today's gigantic cosmos.

What comes out is what originally went in.

An atomic explosion releases a tremendous amount of light energy from an atom. Therein, Trillion Theory says that, 'What comes out is what originally went in.'

Only one material supplied – now build us a cosmos!

Think like an inventor. If assigned, how would you build a universe? Of course, incredible science would be required. But, simplicity might also be an important factor. And, the cleverness of disguising building methods might certainly be a strategic ploy. Building-in a feature such as perpetual spin would be one way to give the universe inner movement and recycling abilities, so that it could grow of its own accord.

Two paramount inventions would be an all-encompassing

building material and then a powerful engine to make the entire system function and operate.

Trillion Theory states that the building material deployed in the building of our cosmos was light. It is an absolutely astonishing material possessing a vast array of incredible properties. TT puts forth light as the material solely used in building our entire cosmos. Every star, planet, moon, solar system, and galaxy built from one material, namely light.

Our universe's design was quite the gargantuan task, by whoever did it. Incredibly this was accomplished through the deployment of just one basic indestructible material, namely a material we see each and every day, namely light.

TT maintains its Light First Theory. Light existed before any matter or stars. TT maintains that all cosmic spheres came about as the result of light being spun into matter.

Furthermore, the engines which spun light into atoms of matter were black holes - the sphere builders.

Properties of light in Trillion Theory.

Light is the most fantastic amazing material in our entire universe. Yet, this light seems to be extremely humble. It has already taken scientists hundreds of years to unveil light's hidden powers; and there is still more to discover in order to understand the total properties of light.

Light is extraordinarily versatile, able to act like a particle at one moment and a wave the next. Light can bend as we evidence by a rainbow. Furthermore, TT maintains that light can bend to such an extent that it can spin into matter. This never happens around us because it takes the incredible rotational speed of a black hole to spin light into matter.

Trillion Theory properties credited to LIGHT:

♦ Light is thus far the fastest cosmic traveler. It speeds at its constant, referred to as 'c' at 186,282 miles per second in a vacuum or in space; 5.9 trillion miles in a single light year; one parsec in 3.26 light years. But, light's speed is slowed when it passes through thicker medium which bend it.

♦ Light, traveling at its top speed, carries its own tool box: electromagnetic spectrum; heat; weight; time.

♦ Light's motto: *Always be prepared.*

♦ Light, it is ever-ready to be spun into matter.

♦ Light's speed stays at its high constant unless slowed by a gravity force. Light slows and bends when it encounters the powerful gravitational attraction from a naked black hole which attracts, bends, and spins light into matter.

♦ Light's speed also slows when it encounters a sphere with a black hole at its core, even though the full black hole can no longer spin light into matter.

♦ Light is thus the universal supplier, the building material, for making everything known as matter in our cosmos.

♦ Light is totally indestructible, thus the ultimate recycling material, recycling on to infinity. Straight line light can be attracted and spun into matter by a black hole and then at the end of billions of years in captivity, that light can escape an atom returning to straight line light.

Light meets the gravity of a naked black hole.

The naked black holes of our cosmos are known in TT as *Gravitational Stations.* Here, light slows well below its normal constant 'c' traveling speed of 186,282 miles per second.

When light encounters a powerful naked spinning black hole, light is attracted and pulled inwards. Once captured, that light is bent, spun, and assimilated into atoms. This light loses more and more speed as each action occurs.

Laws of Light when encountering the ferocious spinning speed of a naked black hole:

- Light slows more as it is grabbed by the naked black hole's strong ferocious spinning speed.
- Light bends to the extreme, coiling and spinning.
- Light spins forming the body of an atom.
- Light spun into matter continues to spin for eons.
- Light, once encased inside an atom, can spin with varying strand lengths to form a nucleus and its electrons.
- Light with varying lengths and thickness of strands form different atoms of elements.
- Light's weight property is provided to an atom of matter.
- Light spun into an atom of matter at the interior of the black hole continues to spin as an atom and will be locked away in this interior jail for billions of years.
- Light spun into more atoms will form the body of matter surrounding the black hole. A small solid sphere is born.
- Light attracted by the gravitational pull from the interior black hole adds to the exterior sphere's surface forming it into a moon or planet. Rotational speed becomes sluggish.
- Light approaching a filled-up sphere with a cloaked black hole can only attract, but no longer spin light into matter.
- Light trapped as matter always wants to escape.
- Light trapped as matter at the interior of the black hole's core is always the first to unravel. With nowhere to go this unspun light is trapped. This matter will heat up the core.
- Light is always successful in its escape from matter, even if it takes billions of years.

Light has a formula of escape: $mc^2=E$. The total amount of energy (E) released from a single atom of matter depends upon its mass (m) multiplied times the speed of light (c) squared (what comes out is what went in). That formula tells of the tremendous amount of light entrapped within a single atom of matter waiting for eons like some genie to escape.

Light is the absolute inexhaustible indestructible material.
Light meets the gravity of a planet or moon (spheres).

If light's encounter is with a filled up sphere such as a planet or moon, that light is attracted towards the sphere's surface by the gravity of the black hole at the sphere's core. That gravity pull goes from the black hole right through the entire matter of the sphere and extends out to surrounding space. The hard matter of the exterior of the planet prevents the already full black hole at the sphere's core from getting its hands (so-to-speak) in direct contact with the light and spinning it into more matter.

Light traveling into such a filled up sphere immediately opens its tool box and sets to work providing light and heat. Light, 100% of its time, is constantly being a universal supplier. Light's other motto: *Be in constant motion.*
Everywhere, these TT Laws of Light are relevant.

We weren't around for the origin of our cosmos a trillion years ago. We weren't even around for the beginning of the present 67th cycle of our cosmos when our present planet and solar system formed billions of years ago. In neither case did we get to see what material was used to construct the many atoms of matter of basic elements. We missed out on that monumental manufacturing process.

Because of that, we live on a planet where everything has

already been spun into atoms of matter. All we do is take the already manufactured atoms and use them in thousands of ways by mixing and matching elements into compounds and even macromolecules (polymers) for plastics.

But, the main point here is that on Earth we missed out on seeing the initial creation stage where light energy was spun into atoms of matter. Now, we mostly see the locked stage (light locked up and imprisoned inside of atoms).

However, on Earth we have been provided with some end-clues or partial end-phases such as fire, lightning or a horrific nuclear explosion where varying amounts of light escape their atoms. Every day we get to witness an end phase when light escapes from the sun. That light departs an unraveling helium atom on the sun's surface and escapes back to freedom as straight line light. That light will either continue to travel for millions of years through space, or meet up with a waiting naked black hole only to be captured and respun into matter once again (new creation phase).

The real magic was how science could have invented and designed the most incredible material (namely light with a host of properties) and powerful machines (namely black holes with engine-like specifics to create mega spin). This inventiveness combined the efforts of both light and black holes to work together sometimes as friends, or sometimes as foes, to make the most beautiful universe imaginable.

An analogy from TT of light being spun into matter.

When it comes to using light to make an atom of matter, a neat analogy is to imagine a grandmother's knitting. Think of light as yarn. Then, notice the beautiful knitting as the lady makes many wonderful garments from her yarn.

She uses different amounts of colorful yarn, various types of knots, a variety of patterns, and different tightness of loops, to create a warm woolly scarf, toque, sweater, mitts, socks, or a blanket. They all look different, but underneath they are all made from one material, they are all just yarn.

Then in reverse, the grandmother can take every one of the knitted garments and unravel them back to their original yarn material. Regardless of any disguise, the garments are all just yarn and can be unraveled back to yarn.

Therein, following on that theme, when you unravel all atoms of matter you discover tons of escaping light (along with the properties of light). To take that thought further, when you unravel all the atoms of a moon or planet or star, you have an absolute incredible amount of light energy released from captivity. Formula $MC^2=E$ comes into play. The amount of energy volume which escapes back to straight line light would be staggering to a zillion zeroes. And going back to the creation phase, the amount of light spun into matter to originally make a planet was uncountable.

Trillion Theory realizes that it may be difficult to accept that all matter is made from one and only one substance, namely light, when your eyes see such an array of matter.

But, light has unique properties which allow it to be spun into a whole number of different atoms of specific types. There exists on Earth over 100 discovered pure elements of matter, and every one of these elements has a specific atom type. There may even be other elements on other planets or in other solar systems or galaxies with unique atomic arrangements yet to be discovered. Each element atom is unique from every other element.

Far-fetched.

At this stage, as a reader you may have reservations as to how these things in our cosmos could be; how light could be the one and only material making up everything, or how black holes spun light into matter and built the spheres, solar systems, and galaxies of our cosmos. You may say, "It's too far-fetched and improbable that our cosmos grew larger from its meager beginnings a trillion years ago to its present colossal size and sphere population. Where's the proof?"

A reply from Trillion Theory (TT).

Is Trillion Theory any more far-fetched than the Big Bang? Big Bang fails tests, doing nothing to explain many things about our cosmos. Whereas, TT attacks these mysteries. In later chapters, TT will discuss ways whereby astronomers can scour certain galactic locations for evidence to prove TT.

Interviewer: "It seems hard to believe that all different types of atoms and elements are made from one material."

Author: "Agreed, your eye sees a wide array of all the various wondrous types of atoms of matter. But, going back to the grandmother, where she uses yarn to make different items such as sweaters, scarves, mittens, blankets, they're all made from only yarn. But the yarn can vary in color, thickness, length, type of loop, and so on. Likewise, light has such versatility spinning into an array of atoms. Every one of the 118 known elements is a pure atom with its own specific atomic arrangement of protons, neutrons and electrons. Yet all the elements have similarities, all supplied by the same material, light. There is also a very big chance that there exist fantastic elements which we have yet to discover."

Interviewer: "And here I thought light was simple."

CHAPTER 6
MACHINE-LIKE
BLACK HOLES

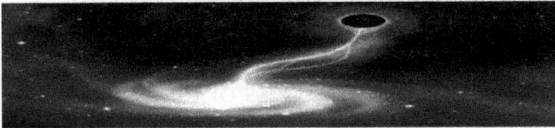

H ow machine-like black holes accomplished the tremendous task of building cosmic spheres. This was done (and is presently being done) by naked black holes spinning light into matter on such a large scale per black hole that an entire bodily sphere such as a moon or planet forms around the spinning black hole. This taming of the light commodity into matter is incredible.

How this began.

Remember the *Light First Theory*, where light is credited as the singular building material of our cosmos, and where it preceded everything else in our universe; a waiting ocean of light, zeptillions of strands of piled light in all directions, available as an ever-supply of energy for our cosmos.

Then, at the origin of our cosmos a trillion years ago, the catalyst engine to begin the process of spinning light into matter occurred with the casting of the first black hole into the static ocean of light. This singular event began a growth process of our cosmos which has been growing ever since.

A singular Black Hole appears in the ocean of light at the origin of our cosmos. It contained incredible spinning power.

With the introduction of the one initial naked black hole into the ocean of light, it did the duty it was programmed for. Deploying incredible spinning speed, the naked black hole broke light free from the static ocean of light, attracted it, pulled it in, and then spun the light into matter.

Had we been there to witness that memorable scene, we would have seen the blackness of the core of the black hole with the brilliance of light rushing into it. The black hole would have used its spinning power to crash against the static ocean of light, breaking away chunks of frozen light, freeing them to release and begin moving as free straight line light, but not for long. Those free rays of light would have been quickly and methodically devoured and spun into atoms of matter inside the belly of the black hole.

As the singular black hole ate from the static ocean of Light, it spun light into matter around itself. Light brought along its tool chest of properties including heat and weight. Empty weightless linear space was left behind as gapes in the static ocean of light.

As it spun more light into matter, the black hole filled its belly full of matter first and then begun to increase the size of its surrounding ball of matter. This was the first planetary sphere occupant of our universe. Around that ball of matter, between the ball and the ocean of light, space formed as the empty spacious byproduct left behind when huge amounts of light consumed by the black hole exited that area.

This empty space formed around the black hole was linear meaning the black hole ate light along an equatorial plane.

A black hole is sphere shaped. When it spins on its north-south axis the gravitational pull created holds matter on the surface. But, outside the black hole the gravitational pull is on a linear horizontal plane which extends out past the equator where it can hold other spheres in orbit. (Like Saturn and its rings).

This linear result comes about because of the spinning action of the black hole on its axis. The gravitational pull caused by the spinning black hole projects outwards from the black hole on a plane horizontal to its equator. This linear flat effect is demonstratively seen in the orbit of the planets in our solar system around our sun. This is also seen in the disc-like shape of the rings spinning around the planet Saturn; and with the disc shape is a spiral galaxy; and with the shape of our entire linear, not spherical, cosmos.

Once that linear oval shaped plane of space built around the extended equator of the black hole, the black hole had to move through space to keep close to the ocean of light.

One might ask why more light wouldn't have just flowed from the ocean of light into the now empty space. The answer is that the light ocean is static, comparable to frozen. The black hole must loosen static strands to free-flow them.

As this process continued on, the black hole grew more matter around itself, growing into a sphere. Space around the black hole grew larger as more light was taken from the ocean. The size of our cosmos increased. The static frozen-like ocean of light still occupied the outer perimeter, while inner space became a larger linear-shaped empty void.

Thereafter, as more black holes ate from the ocean of light, many more spheres were built congregating into solar systems. The congregation into solar systems was due to the gravitational forces supplied by the black holes inside of the spheres. These black holes used the force of their gravity to hold other smaller spheres in orbit.

Till Trillion Theory, no one knew.

Till Trillion Theory, black holes were never thought of as having a definite duty in our universe. Till now, No one knew their true purpose as builders of the spheres of our cosmos.

Till Trillion Theory, a black hole was never thought to be at the center of every moon, planet and star.

Till TT, no one ever surmised that light eventually escapes from a black hole after being trapped for billions of years.

Till TT, no one theorized how the eventual escape by light from a black hole is part of the sphere recycling process.

Till TT, no one had theorized that black holes are actually alive in their own way; with an animalistic-like appetite to devour light and hold it tight; to become the largest and strongest capable of dominating smaller spheres; able to survive and replicate through the ages; and able to be the dominant engines and builders of our cosmos.

Till TT, no one had theorized that there was an entity at the core of Earth and every sphere (moon, planet, or sun) in our cosmos. This entity doesn't die when Earth ends. Rather a naked black hole survives and splits during fission such that two new naked black holes begin the next 15 billion year cycle of a new solar system. Black hole survival and replication, and an oceanic supply of light are key reasons our cosmos has grown to over 73 quintillion stars.

Albert Einstein laid the theoretical foundation for the existence of black holes by predicting that light would bend when nearing a sphere's gravity. However, no one till TT had surmised that black holes spin light into matter; or, surmised the existence of a black hole living for billions of years at the core of every sphere (moon, planet, star).

Only on rare occasions do astronomers get to telescope a naked black hole wherein it has shed its outer sphere, and is at that precise moment attracting and devouring new light to erect a new sphere around itself.

Because most of their time is spent hidden away, black holes were never fully understood. Humans had no original inkling, till TT, that black holes were related to the creation of spheres and to the recycling process of our cosmos. Rather, black holes were simply thought to be rogues in space where light disappeared into a pit which possessed phenomenal gravitational forces, never to return.

The monumental discovery of black holes by astronomers was just a first step in uncovering their secrets: Till TT, astronomers have incorrectly thought of black holes simply as invaders, looking to devour a star, for no good apparent reason. Now, TT gives black holes a whole new importance, making them prevalent throughout cosmos as instrumental centers of birth and rebirth of every cosmic sphere.

Just like $E=mc^2$, where both energy and matter cannot be destroyed but only interchanged, black holes are master survivors and recyclers. A black hole can never be destroyed even by the force of a Supernova. Try to destroy a black hole and it simply splits replicating into two new black holes.

Over the last trillion years, black holes have been *sphere factories* hard at work building and recycling the spheres of our cosmos. A basic in TT is how our cosmos originated and grew from black holes being recycling sphere factories.

Initial Conclusions:

Black holes are the rotor engine mechanisms which spin light into matter around themselves.

This tremendous scientific-type invention of both of the partners, light and black holes, is absolutely extraordinary. The tool box within light and the powerful spin mechanism of engine-like black holes were both inventions made to last and endure forever. It was pure genius that cosmic origin is the result of black holes devouring light. Also it's genius how the interaction of light and black holes grew our cosmos.

The various masks of a black hole:

A black hole gives different looks depending upon the amount of light it has or hasn't spun into matter around itself. Naked black holes appear just as the words describe, naked and empty as they are just beginning their search for light. This is the best time see a black hole's components.

However, once a black hole has spun sufficient matter, it can hide inside the core of the sphere of matter it has spun. These cloaked black holes can be hiding behind a number of masks or disguises, as a moon, a planet, a star. Black holes give different looks, dependent upon the stage in their cycle.

A black hole has major phases as it does recycling.

Recycling of a sphere follows a sequence of major phases for one cycle. This one cycle goes from a naked black hole into a sphere and then recycles back to a black hole. One entire cycle usually has a 13-15 billion year upper limit.

During a 15 billion year cycle, a black hole goes through phases. Example: Earth is in a mid-cycle planetary phase, so the black hole at its core is masked behind a hard exterior coating. Whereas, our sun is in a late-cycle star phase, so the black hole at its core is masked by the fiery exterior. Note: A cycle can be shorter than 13 billion years if a star goes Supernova ahead of schedule and destroys its solar system.

At the beginning of each new cycle, a black hole always starts out with a naked phase. But as it consumes plentiful light which it spins into matter, the black hole masks itself inside of a sphere such that we only see its outer appearance in the appearance of a moon or planet, or as a sun.

Phase One – Naked Acquisition and Creation of Matter

We see a black hole spinning light into matter.

There is a supply of light rays, either from the static ocean of light, or as free straight line light traveling through space.

Phase one, the naked acquisition of light and creation of matter, always begins with a dark empty naked black hole beginning to spin light into matter. Naked meaning that the black hole is empty and has not yet spun light into matter around itself. This is the type of naked black hole introduced into the static ocean of light at the origin of our cosmos; or a black hole surviving when a star goes Supernova.

The first assignment of any naked black hole is to attract, capture, spin, and entrap light into extremely tight dense matter. In this CREATION OF MATTER phase, the black hole appears as ebony as black can be. However, brightness can occur when a stream of light gushes into the black hole.

Phase one takes a short time frame. 'Blink and you might miss it.' Duration depends upon the amount of light directly available for the black hole to spin into matter. If the black hole is close in proximity to the static ocean of light, it can eat and fill up quickly within just years. If far from the ocean of light, and having to fight against other naked black holes for light, the filling process might take hundreds of years.

Creation of Matter phase:
a black hole spins light into matter in and around itself.

The black hole has more than just the ability to spin light into matter in its core. It has the power to use its tentacles to continue to spin light into matter and add that matter in layers around itself. These layers build up and become the outer surface of a sphere forming around the hole.

Phase Two – Lockup Control

Lockup occurs once the black hole becomes full. This total lockup control phase can last 5-10 billion years. We see a small sphere as a moon or planet.

The lockup control phase follows the initial spinning of light into matter phase. The teen-aged black hole has spun light into matter so that the size of the ball formed around it increased dramatically in size. Once a black hole has spun sufficient matter it no longer appears as a naked black hole. Now, the black hole is cloaked inside a sphere.

Adding more matter around itself, the black hole appears as a moon or planet. Size of this sphere is directly related to the black hole's size and the amount of light spun to matter.

Once this maxed-out size is achieved and the black hole has finally spun as much matter as it can for the size of its black hole, duties for the black hole shift to lockup and control. All effort now switches to the new task of control (LOCKUP), trying to prevent matter from unraveling to light.

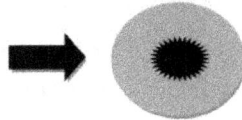

A black hole locks up light as atoms of matter around itself.

The many layers of matter comprising the main body of the moon or planet are held in place by the strong gravity hold of the black hole's helix, its core compartments, and tentacle filaments. Everything is held in place by the force created from the power spin of the black hole. This gravity extends beyond the surface of the sphere to attract and curve more light inward towards the sphere, and extends far enough outwards to hold any other smaller spheres in orbit.

While the black hole still wants to spin more light into matter that task becomes ever more difficult as it fills up. Each iota of matter which the black hole adds to its exterior surface blocks its ability to spin new light into matter. It can still attract passing light to its surface, but no longer can the black hole spin by-passing light into new matter.

Eventually, the spinning speed of this fuller black hole begins to decelerate. The rate of spin at the Spiral Helix of the black hole decreases as it becomes sluggish from all the matter filling all its compartments and its surface. Light, can still be attracted, but is simply absorbed or even reflected.

The black hole now switches its focus from spinning more light to the task of control over the zillions of atoms which it has already spun into matter; trying to prevent the spun atoms of matter from unraveling back to straight line light.

Phase three – Loosening and some Escape

Lockup tries to continue even while LOOSENING begins (we see a planet with a more active surface). Once this loosening gets more rampant, ESCAPE commences. This phase can occur as early as 5 billion years and last 10 billion years.

After years of first spinning light into matter, followed by billions of years of holding matter in place, the moon/planet begins to weaken. Because, the black hole at the core of the sphere always overate by over-consuming light, a day comes when it must pay for having to maintain total control over too large an amount of light spun into matter.

We examine how a black hole tries to maintain control over the matter which it spun from light, and how a loss of some control results in the beginning of a loosening phase.

When the black hole fills from its eating, it still attempts to draw more light to its surface and tries to devour that light and spin it into matter, but it can't because the matter on its surface blocks the way. An example would be Earth.

Inside Earth, our black hole resident has strong control over surface atoms. But with this strong attempt to hold surface matter and keep our moon in orbit, and attract light to the surface, our black hole has lost control over atoms deep inside its core. Those oldest spun atoms are the first to begin the process of LOOSENING. The nuclei of these oldest atoms now begin to loosen as they spin less tightly. They expand and unspin in their attempt at ESCAPE. However, for loosened atoms there is nowhere to go as they are buried deep below miles of matter. Yet, this loosening creates heat and expansion inside the core of the black hole fires up.

The planet around the black hole grows larger. The black hole inside the sphere is now small in proportionate size.

We can only imagine the turmoil for the black hole as a hot fire expands deep within its belly, commencing years of losing control. That fire inside the black hole's belly expands, placing tremendous pressure on the structure of the sphere.

Fissures form as lava created from the internal fire fiercely expands outwards searching for an escape route towards the surface. With fissures are blocked by solid matter, something has to give as the surface undergoes dramatic change.

The internal pressure can force areas which are under the surface to buckle projecting large masses of matter upwards (mountains form). Also, surface expanded continental drift occurs. Eventually, lava breaks through as a historical first volcano (Pimple Burst) explodes gigantically to relieve the pressure. Volcanic activity for the sphere will peak. Lava which reaches the surface quickly cools to form rock.

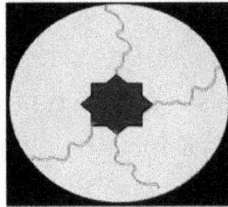

The oldest spun matter at the core of the black hole releases first and unspins to form a fire which searches for a way out through fissures to the surface to form the lava of volcanoes. Once this process begins it is irreversible. It may take billions of years, but in the end a fireball star is the final result.

If the planet survives long enough, lava eventually engulfs the entire surface transforming the sphere into a fireball star.

This loosening phase of a moon or planet could include 100's of sub-phases over a 5-10 billion year period. In our solar system, although all spheres are similar in age, their state (solid, or gaseous, or plasma) is dependent upon the size of their black hole, and how much that black hole overate, and how well that black hole maintains control.

Earth (a solid surface) has a hot interior where the escape of atoms to the surface has been going on for millions of years. It will be many more billions of years before Earth can loosen more and have a more gaseous liquid surface.

Jupiter, on the other hand, the largest planet in our solar system, really saw its black hole overeat during construction, so its interior is extremely hot. This heat has turned the surface into a more gaseous liquid state. Trillion Theory says that had it not been for our sun, Jupiter would have been the top candidate to become the sun of our solar system.

Sun has the largest core black hole in our solar system. It ate by far the most light when our solar system was forming. Sun was glutinous as to how it pulled in light and spun it into matter. Our sun, with its size dominance, formed the biggest body and also gained central status as the most prominent gravitational force holding all other lesser sized spheres (planets and moons) in sun's orbits. Our sun was also the first of all the spheres in our solar system to loosen control over its matter. Fire started early in its belly, flowing uncontrolled to the surface, which is now totally plasma. Atoms now easily unspin and release as light free to travel through space, or until meeting another gravity station.

Phase four – Supernova and Obliteration

Occurs at around age 10-15 billion years, towards the end of a star's life. For a sun, the unspinning of atoms becomes rampant as the sun has solar flares and emits tons of light.

Phase four deals with the final phase in the life cycle of a sun. Very few spheres reach star status, as those of lesser size are destroyed when their sun goes Supernova.

In phase four, the sun loses control of its contents. The sun of our solar system is in the mid-stages of phase four. It has expanded, growing ever larger. Fatigued, its contents loosen, each atom becoming larger thereby expanding the suns diameter. Light departs from its surface at a hectic pace; freeing atoms back to straight line light.

Hot atoms are now prevalent from the black hole at the sun's core right out to its outer plasma surface, where solar flares leap from the overall intensity. A fireball is the result.

In the end, the star known as our sun will expand and eventually die exhausting all its fuel. When our sun expands gigantically, dies and goes SUPERNOVA, the expanse of this explosion will be far reaching. The Supernova will destroy some or all the planets and moons within our solar system depending upon a sphere's proximity distance from our sun. Prolific carnage of a solar system by a sun - OBLITERATION.

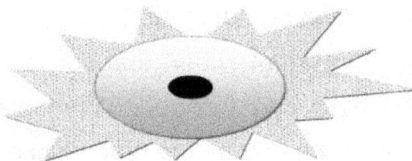

At the point of total fatigue, the black hole at the core of the sun loses control of all its spun matter as the sun's contents explode going Supernova and the sun dies.

Imagine the blast from an uncountable number of atomic explosions occurring simultaneously when any sun goes Supernova. The heat from all the instantaneously escaping light melts all of the spheres (planets and moons) in that sun's solar system. This Obliteration destroys all the surfaces, leaving naked black holes from the core of every sphere as the only survivors. Additionally, each sphere in that solar system will experience a flash Supernova of its own melted away body, thereby increasing Supernova's overall power.

After the Supernova and then the obliteration of spheres in that solar system, tons and tons of pent-up light escapes from the matter which had been locked up on and within those spheres. That freed light departs, free to travel space, or until called upon to once again spin back into matter.

However, much of that happy freed light only experiences temporary freedom, as the now naked black holes which survived the Obliteration commence their battle against one another to spin that light into new atoms of matter for the building of the bodies of their new spheres. The creation and the building of a new solar system commences once again.

Phase five – Replication

After Supernova, a new startup phase occurs. This startup may only take a few short years. There exists a graveyard of naked black holes which survived the Supernova and the Obliteration. Now, resident within that space are double the number of replicated and twinned naked black holes. For every old black hole we now see two new naked black holes.

When the star gigantically expanded to go Supernova, all of its contents of spun matter changed back to light which escaped from its surface as straight line light. Surviving was the now empty naked black hole which was at the sun's core. This pliable elastic black hole can never be destroyed.

At this point, this now abandoned black hole regains its prolific spin speed. With the terrific unraveling force of a Supernova, the now exposed black hole spins furiously in the opposite direction (for every action there is an equal and opposite reaction). From the violence of the Supernova, and a ferocious counterspin, this reverse force splits the helix of the black hole in two, splitting the black hole's core. The two split parts recoil and repel away from each other. Two new duplicated black holes replace the original; REPRODUCTION, now two black holes instead of one. Replication is the end-phase of the old cycle; now begins the next creation phase.

Immediately these two new black holes go back to their virgin instincts and begin competing to devour light. Their first use of the new light they spin into matter is to fill their gut and then to increase their size by the doubling of their compartments via fission.

During Obliteration, the destroyed moons and planets in the solar system explode instantly when their central sun's Supernova destroys them. While their emancipated bodies have returned to straight line light, each black hole remnant which was at the center core of each destroyed sphere survives obliteration (black holes can never be destroyed). Each black hole splits in two from the force of Obliteration.

The end reproduction result is the doubling of black hole numbers heading into the next cycle. After Obliteration, the area around all the new naked black holes becomes a war zone as all the new naked black holes cannibalistically fight over and snap stands of light to spin that light into matter.

The black hole which was at the center of a destroyed
sphere splits to form two new black holes.

How to view cosmic phases.

Unfortunately, because of short human lives relative to
cosmic time, our view is limited. Of the 5 phases, we see
mostly the lockup stage on planet Earth. However, powerful
telescopes can show us Supernova and Creation phases.

After the Supernova of a sun, all the planets and moons
of that solar system are torched by the exploding sun. Each
sphere, during Obliteration, faces its own explosion.

After Obliteration occurs, naked split black holes (double
the number after Replication) are the graveyard survivors of
all the supernovae which occurred in that particular solar
system. That graveyard will only be visible for a short period
as all naked black holes in the graveyard will begin their
battle to spin light into matter. One new large solar system
or two adjacent solar systems can be the end result.

For we humans to see the precise creation process within
a black hole is virtually impossible, even with the advantage
of powerful telescopes. Someday, it might be possible. Such
a view would do much to prove Trillion Theory.

Summary of light and black holes.

Take away light and black holes, nothing in our cosmos
would be possible. Both are incredible scientific inventions.

Light travels through space for billions of years, or until
influenced by a gravity force. Light, the universal carrier, is
loaded full of supplies necessary to create new matter.

However, light is not a Merlin Magician as it cannot create matter all on its own - it requires an engine to spin it. The powerhouse to capture and engineer light into matter is the black hole. Black Holes are our cosmic engines of spin.

As light travels near a black hole, here are the particular forces at work: The black hole contains a spinning helix axis with fantastic rotational speed. As light passes closely by, it's attracted and pulled in. This captured light is then curved such that the head of the light ray turns sideways and spins like a barber's pole thus creating a spinning helix which pulls and spins the rest of the tail of the ray in behind itself. The result is a spun light ray now in motion as an atom.

All the properties that light possesses are now available from its tool box: spinning motion; heat; weight; length of strands; thickness of stands; elasticity of strands; a head for the nucleus of an atom; a tail for electrons.

All follow a perfected design pattern for the type of atom element being constructed. This is a tug a war between the light and the spinning black hole. Light wants to stay free to travel, but it is overwhelmed by the gravity of the black hole. During this struggle light is pulled, elongated, twisted, and even snapped into lengths in determining the type of atom.

The weight which light possesses provides 99.99% of the weight and mass which is found in all cosmic spheres.

The many special properties of light allow it to be the ultimate recycling substance, going on to infinity, from light to matter, and then matter back to light. Cosmos continues to grow and expand, being self-perpetuating. "Nobody has to turn the crank when a part of our cosmos has to recycle."

The properties of light are many and unusual:

▶ Light is totally indestructible, regardless of how many times it recycles.

▶ Light can be sucked in and spun into matter by a powerful naked black hole.

▶ Light can spin sideways when in contact with a black hole's helix.

▶ Light can form the nucleus/electrons of an atom.

▶ Light spun into matter possesses the propensity to continue to spin as an atom for eons of time.

▶ Light possesses weight, providing weight for matter.

▶ Light trapped as an atom of matter always wants to someday escape back to its straight line form.

▶ Light will always ultimately escape.

▶ Light carries the property of time within itself.

Light spun into an atom spins as an atom for eons.

Atoms spin for eons because each atom is made from light. Light's motto is that it must be ever in motion. (Light is only motionless initially in the frozen static ocean of light).

Now, one might think the light might come to a grinding halt when a black hole spins it into matter, but not so. Light continues to spin round and round for eons inside the atom.

Also, let's look at the make-up of light. We think of light as this thin long filament whisking forever across space. But, a ray or beam of light does have a head and tail. The light moves in pulses, like waves, as the elasticity of light pulls back to recoil and then pushes forward to keep ahead again. The wave is like a linear pump allowing the ray of light to pulse forward as it travels. When captured inside an atom, light will continue to use its elasticity.

In the atom, this central axis helix rod within the nucleus is screw shaped and has an elastic property causing it to pump up and down, alternately shorter and longer, which allows it to transfer this up-down motion into a fast spinning motion causing the body of the atom (electrons) to spin around the helix. The atom's built-in elasticity allows the pumping motion to continue unabated for eons.

The cosmos is very similar right from the tiny atom to the monster-sized black hole at the center of a spiral galaxy. All utilize a similar pumping action to sustain continual rotation.

A closer look at the power engine of black holes.

Black Holes are extraordinary. Just imagine the force of attraction required to spin humongous amounts of light into tightly packed atoms such that trapped light cannot escape those tiny spinning confines for billions of years.

Black Hole

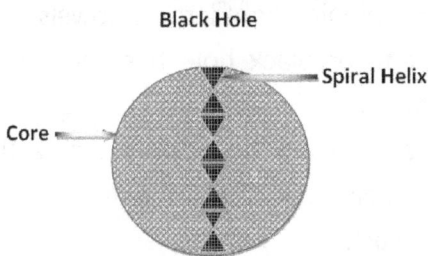

Component parts of a black hole which have to do with spin: interior spiral helix axis is the engine of spin.

After black holes spin light into matter, they deploy that extra added body weight and mass into a stronger gravity.

Black holes are indeed without equal. So, what is it about a black hole which allows it such power in our cosmos?

- A ferocious spinning mechanism.
- A pump that keeps the spin going.

The naked black hole axis utilizes an up-down pumping action to create its tremendous bodily spin of thousands of revolutions per nanosecond. This creates a gravity zone. As nearby light approaches, it is lured and spun into matter.

The head of the light ray turns sideways and spins like a barber's pole thus creating a spinning helix which pulls and spins the rest of the tail of the light ray in behind itself. The trapped light ray is still in constant motion as an atom but now spinning below light's normal speed. This ball is tightly spun, occupying only about a 1,000,000,000,000th of the area previously occupied by the light. All the properties of the strands of light are now within the black hole. While space becomes a vast empty weightlessness which light vacated.

A look at the spheres which a black hole can build:

- A **naked black hole** is in the absolute beginning process of capturing and spinning light into its bowels.

- A **somewhat fuller black hole** has spun more light into matter to form a ball of matter around itself.

- A **full extra-small black hole** appears as a tightly packed moon (extra-small black hole at the moon's core).

- A **full small black hole** appears as a tightly packed small planet (small black hole at the planet's core).

- A **full large black hole** appears as a gaseous less tightly packed large planet (large black hole at large planet core).

- A **full extra-large black hole** in a solar system will form the largest planet, holding smaller spheres in orbit. This extra-large black hole will have the loosest atoms, and as it ages, weakens and tires, its surface changes from a gaseous state to a fire plasma state as it becomes that solar system's sun.

- A **full supermassive black hole** at the central hub of a galaxy plays by a higher-level set of rules and cosmic laws. These parameters will be presented in later chapters.

Interviewer: "I'm fascinated by the ocean of light. Where did it come from? Also, from where do black holes originate?"

Author: "Those are huge questions. To know for certain one would need to rap the door of the scientist(s), or Artisan(s), or creator(s) of this universe (if such entities exist), and make that query. Whatever the answers, scientific knowledge (far beyond ours) was deployed in the inventions of both light and black holes. Also, there was a clever strategy deployed in the methods of operation. As to who or what supplied the materials, I've always loved a mystery. Something has been cleverly egging us on, tossing us tidbits while saying "look at lightning, nuclear explosions and supernovae as clues."

Interviewer: "How big is this ocean of light?"

Author: "Endless. This is a difficult concept to grasp because we always think in terms of there being a start and an end, a border or edge, to anything physical. Instead, think infinite."

Interviewer: "Hold on Ed. Why doesn't light from the outer ocean of light just move inwards and fill empty space?"

Author: "Because in 2013, lab scientists discovered a brand new property of light thereby supporting my theories: Light was frozen (so-to-speak) as it stopped moving. Similarly, the entire ocean of light is a static (frozen-like) resource. A ray of light has to be loosened from the frozen ocean by the spinning black hole. Once loosened from the light ocean, the ray can attain the speed of light. But most often, before the ray can escape the scene, it is attracted, slowed, curved, and then spun and locked away for eons as an atom of matter."

CHAPTER 7
BLACK HOLE TO
PLANET TO STAR

Theoretically, every black hole which spins a sphere of matter around itself should be destined to live its long 15 billion year life cycle from black hole to sun. But in reality, very few spheres possess the wherewithal to survive long enough to complete such a task.

Some suns see their lifespan cut shorter when they are attacked by a rogue black hole. And, the hard-body moons and planets of a solar system can face early curtailment when the sun of their solar system goes Supernova and obliterates them. They speed from hard-body, melting from the heat of the Supernova, and they too explode.

Nonetheless, this chapter follows the successful lifespan cycle of a sun from its start as a naked empty black hole, evolving to a hard-body large planet, and finally evolving to the central orb of a solar system as a sun, ending finally with its death 13-15 billion years or so into the cycle.

Here are the major phases in the cycle of the sphere body created around a black hole: short duration beginning birth (CREATION); long midlife with locked up atoms (LOCKUP); a long slow unraveling as we see with our light emitting sun (ESCAPE); sun explosion (SUPERNOVA) and (OBLITERATION) of its solar system; and a rebirth (REPRODUCTION). The end of one 15 billion year cycle quickly ushers in the beginning of the next cycle with double the number of black holes.

Durations of these various phases denoted by >.

CREATION > (short phase).

LOCKUP >(longest phase).

ESCAPE > > > (short-medium length phase).

SUPERNOVA > (short).

REPLICATION >(short phase).

Following is the 17 stages of one 15 billion year cycle. These stages bring further detail to the 5 main phases.

Stage 1: Naked black hole spinning light into its bowels.

Stage 2: Black hole spins light to form an exterior body as a small moon or planet.

Stage 3: Fire occurs in the core of the moon or planet where its oldest spun atoms begin to unravel.

Stage 4: The interior fire expands, swelling the sphere. The sphere's surface experiences continental drift.

Stage 5: Earthquakes occur as the lava seeking an escape route pressures the surface and the crust buckles.

Stage 6: Mighty inaugural volcano Pimple Burst occurs.

Stage 7: An Ice Age hits the planet.

Stage 8: Planet re-warms after the Ice Age.

Stage 9: Black hole utilizes gravity to hold things on surface.

Stage 10: More atoms loosen, move upwards; surface thins.

Stage 11: Free flow of lava creates a liquid surface.

Stage 12: The planet emerges into a small star.

Stage 13: Small star expands to a larger star.

Stage 14: The aging star expands extraordinarily.

Stage 15: Supernova. Star dies as it explodes.

Stage 16: Obliteration. Sun destroys its solar system.

Stage 17: Rebirth (Reproduction) (Replication).

Two new back holes are formed for every old black hole.

Each stage of a 15 billion year cycle is further explained: Stage 1: (short stage: elapsed time from hundreds, to thousands, possibly up to one million years).

A naked black hole begins its task with the spinning of matter into its bowels; then, spinning more light into matter to form a ball of matter around itself as a moon or planet.

A naked black hole spins its central axis to attract and spin light into matter. It has incredible spin speed and extremely strong gravitational pull on light. This light spun into matter adds to the black hole's axis length and girth until it is full as a ball of matter. This stage adds weight to the black hole as light carries along with it the property of weight when spun.

Duration of this building process is dependent on: the size of the black hole; the available easy supply of light; and the size of nearby competing naked black holes. Eventually, the ball increases to the size of a small moon or planet. Space, a byproduct, surrounds the sphere as the void empty of light.

This light which the black hole attracts can be from the static ocean of light, or light traveling from supernovae and obliterations, or light traveling from stars. At the origin of our cosmos, in the first 15 billion year cycle, the black hole(s) were all close to the static ocean of light, fill-up came easily. Whereas now, a naked black hole may find itself far removed from the outer perimeter static light ocean.

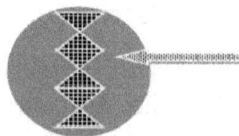

Light is first sucked into the axis and then into the core of the black hole. Once full, more matter is added around the core to form the hard body of the moon or planet sphere.

The sphere now sits in the space it has created around itself with the absence of the light used to build the sphere.

Stage 2: (at 1 million to 2.5 billion years).

Small moon-planet grows larger, takes shape, adds surface.

At this point, the black hole at the core may be 100's of times smaller than the orb it has created around itself. Its core is packed and its surface hard. The surface may display rough features caused by irregularities: wind; temperature extremes; and pit craters caused by meteoric bombardment from rocks cast into space from other spheres.

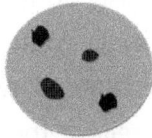

Centrifugal Force of spin causes the sphere to be slightly fatter at the equator. Craters begin to mar the surface struck by meteors fired into space by huge volcanoes know as Pimple Bursts from other spheres.

Stage 3: (at 3 billion years).

Fire in core of black hole begins as atoms unravel.

Every tentacle compartment within the black hole at the core, which spun light into matter, switches from capturing (Guzzling) to control (Holding Gravity). As the interior core of a spheroid, the black hole is now dedicated to control. But, eventually over eons of time, this control loosens in the core, where the oldest spun matter exists. As solid atoms of matter unravel, a tiny fire is slowly ignited deep in the inner core. This matter unravels forming hot molten lava.

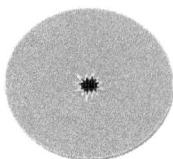

As the sphere becomes right full and increases in size, no more light can be spun into matter. The black hole at the center loses control of its oldest spun light, as matter unravels into a fire.

Stage 4: (at 3.5 billion years).

Interior Fire grows causing further planet expansion.

The interior core fire grows larger when even more atoms unravel deep in the black hole's core. This fire continues to expand moving outwards. Loosened expanded atoms take up more area than previously tight densely spun atoms. This ever expanding fire eventually creates so much pressure that the moon or planet has one of two choices: explode, or otherwise find a way to relieve the pressure. Fissures occur up to the surface from the internal pressure allowing lava from the interior to find escape routes to the surface.

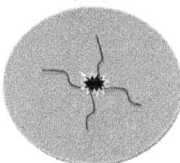

As interior fire grows, pressure mounts causing fissures which the fire can follow to try to get outwards towards to the surface.

Stage 5: (at 3.75 billion years).

Earthquakes occur as the lava pressures the surface.

Trapped lava exerts more pressure causing fissures to become wider and longer to the surface. Internal pressure intensifies as lava crawls up fissures searching for an escape route in order to relieve the pressure. The planet's structure undergoes duress resulting in a large buckling of the surface as gigantic earthquakes occur as parts of matter thrust upwards to form mountain ranges. Also, continental drift occurs on the planet's surface.

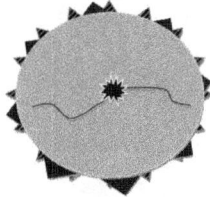

As the lava fissures pressure the surface, gigantic earthquakes cause upheaval as mountains of rock are forced outwards.

Stage 6: (at 3.8 billion years).
The great volcano named Pimple Burst occurs.

Lava, under extreme pressure, becomes trapped in many of the fissures leading to the surface. The entire planet feels ready to explode and destroy itself unless the lava can crack the surface. At the weakest point, lava fires upwards through a wide fissure. The surface of the sphere buckles upwards from the force of the attack. At this point in the planet's history, one of two things occurs. The planet can explode from the internal pressure, or a large amount of aggressive lava finds a wide fissure to monumentally break the surface.

Finally, the initial power-packed volcano named **Pimple Burst** explodes through the surface, packed with so much pent-up lava that the surface juts upwards and the thrust catapults rocks. This first fracture of the sphere's surface is so immensely powerful that rocks are propelled from the surface out past its atmosphere hundreds and thousands of miles into space. They become meteors which can smash into nearby moons/planets to crater the receiver's surface.

Back on the planet, there will never again be a volcano as powerful as Pimple Burst. However, this will be the Age of Volcanoes as many break the surface spewing their lava. New land formations will occur from the lava's pressure. Tons of rock and ash are forced into the atmosphere.

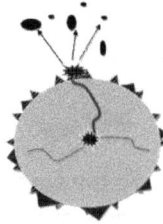

Pimple Burst, the first enormous volcano on the planet swells the entire sphere almost causing it to explode. Pimple Burst is so powerful that it fires rocks like a rocket, overcoming the gravity of the planet sending debris into space.

Stage 7: (at 3.9 billion years).

Ice Age hits the planet.

There may be decades or centuries of cold temperatures right after Pimple Burst as that initial release and Age of Volcanoes spews tons of rock and dust into the atmosphere. While the pressure from the fire at the core has found relief, the surface must face a new challenge as light to the surface is blocked by thick clouds of volcanic ash. This precipitates a prolonged drop in temperature causing the surface of the planet to become frigidly cold. An Ice Age occurs.

After Pimple Burst, the entire sphere experiences a cold Ice Age. An ash cloud blocks out the entry of light.

Stage 8: (at 4 billion years).

The planet re-warms after the Ice Age.

As the cloud cover of ash dissipates, the planet's surface re-warms as light is accepted. Volcanoes become less active; internal pressure from lava pushing upwards on the planet's crust is relieved by lava flowing direct to the surface. The size of the planet has increased like an inflated balloon. As the planet further expands, surface separation occurs as main land masses split apart into continental drift.

As the mushrooming planet expands, Continental Drift occurs.

Stage 9: (at 5 billion years).

Mature planet utilizes gravity to hold surface things.

Determination for the mature planet depends upon its distance from its sun. If it is lucky enough to reside within the Goldilocks Zone, not too close nor too far from a sun, a hospitable primordial soup situation may present itself for life to be seeded. Living organisms on the planet's surface experience the strong holding gravity exerted by the black hole at the core. All upward efforts result in a drop back.

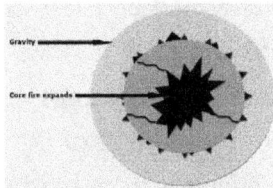

Black hole at the core of planet utilizes gravitational pull to hold matter, even as fire in its core continues expand to the surface.

Stage 10: (at 7 billion years).

The planet's surface thins.

As the central core of the planet loses more of its hold on old matter, and the hot interior fire expands, the interior of the planet's crust is continually depleted and thinned. The surface becomes warmer, then hotter.

The crust of the planet continually thins more from the interior. The overall thickness of the crust is diminished. Anything and everything on the surface feels the increased heat as the internal fire eats away at the tinning crust. The planet sees speeded expansion in size as the interior fire gains in furor. Continental Drift undergoes varying episodes of speeded expansion as continents pull further apart.

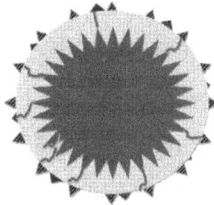

Planet's surface thins as core fire expands reaching the surface.

Stage 11: (at 8 billion years).

Planet's surface becomes more liquid.

The crust of the planet becomes frail and thin as the lava below eats away. Volcanoes no longer need to erupt since the surface of the planet had become so thin lava can easily find an avenue to freedom as it flows freely over a large portion of the surface. Temperatures dramatically increase There is no escape from the heat of the growing lava fields. The melting surface changes from hard rock composition to a soupy liquid. The ground feels like hot coals. Land masses become unnoticeable. Newly spouted lava no longer cools to add to the hard land surface as in the past. Clouds disappear since a water supply simply isn't available to create a vapor cycle between land and sky. Lakes, rivers, and oceans disappear. Eventually, the planet becomes a fire ball.

This planet's liquid stage provides evidence that cosmos works differently than old Big Bang theory. In sharp contrast, TT states that the planet heats up rather than cooling in the cold depths of space. The fire starts, percolates, and grows from the loosening at the core of the black hole. That fire grows outward over billions of years; planet turns fireball.

The planet's surface experiences more of a liquid state as lava from the internal fire encroaches on areas on the surface.

Stage 12: (at 9 billion years).
Planet emerges into a small star.

The crust of the planet disappears as the internal fire melts the entire surface turning it into a fiery state. No longer does lava harden as rock; it now remains hot and molten. The ball has been transformed from a planet, absorbing light and heat, to a small star emitting light.

Planet, with surface fire, emerges into a fiery young star.

Stage 13: (at 10 billion years).

The small star expands to a larger star.

As the fiery star changes all its molten rock and lava from hard matter to a liquid-gaseous state, the star grows in size from a junior to medium sized. Plummets of fiery gaseous eruptions occur from the surface where atoms unravel, escaping to freedom back to straight line light. At this stage, the sun is 4,500 times larger than the black hole at its core.

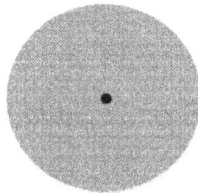

As the black hole at the core of the sun loses control of more atoms, the star expands growing ever larger and hotter. It is now large enough to control a solar system in its orbit.

Stage 14: (at 13 billion years).

The star expands extraordinarily.

The star grows immense in size maturing from a medium to a huge parent star. Age shows as the star displays erratic uncontrolled expansion. Good fortune might allow the star a long stately life. However, the massive star is now vulnerable to a perilous attack from any nearby naked black hole.

More light breaks free after eons in LOCKUP. For a few billion years the sun becomes the solar system's heating house.

Stage 15: (at 15 billion years).

Supernova - the star dies as it explodes.

This elder star grows into a colossal mammoth old star. The oldest stars seem to succumb to old age once they reach 12-13 billion years or so. At the surface of the sun, photosphere helium atoms break apart as light is emitted - freed after billions of years of captivity. Finally, the star loses entire control and explodes providing the universe with one of its brightest sights, Supernova. If there is a rogue naked black hole in the area, the end for the sun may even come sooner. The naked black hole's pull on light will speed the unraveling of the sun, creating a bright Quasar, with a rush of light from the sun into the predator naked black hole.

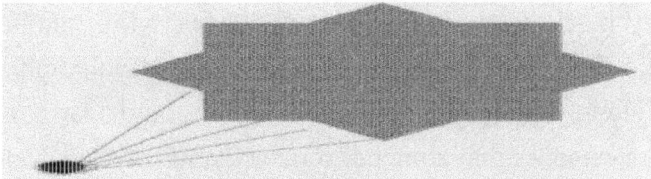

Supernova - the star dies as it explodes.
The surviving black hole can create a brilliant quasar, as a massive rush of light goes back into the black hole.

Stage 16: (at 15 billion years).

Obliteration. The star destroys its solar system.

The sun's enormity soon engulfs the closest planet. Then, Supernova does the rest. Inevitably, humongous Supernova explodes all of the sun's remaining contents at one time.

Planets and moons fracture and face annihilation as the star destroys most of its own solar system. These moons and planets never get to experience slow growth into a star, for at the point of Obliteration their entire matter is exploded like mega-zillions of nuclear blasts. Planets far away from their sun might survive or be thrust further out into space.

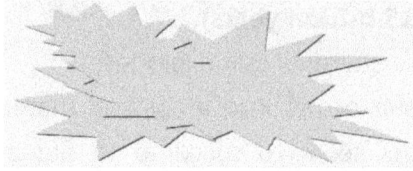

The Supernova power burst from the dying star can destroy
nearly every moon and planet in its solar system. This
destruction of the solar system is known as Obliteration. Only
those further moons or planets furthest away might be spared.

Stage 17: (at 0 years, start of next cycle).
Reproduction: Two new back holes are formed. Replication.

The neutron black hole which was at the core of the
exploded sun, plus all the black holes at the cores of the
annihilated planets and moons, each split into two new twin
binary black holes. In the next cycle, twice the number of
black holes will engage in a raging battle to devour light.

Therein, the cycle is complete in one particular zone of
the universe, as that zone has renewed and recycled itself
yet again. The universe is so vast that many zones unknown
to one another may recycle at the same or different time
periods. Approximately every fifteen billion years, all zones
of the RECYCLIUN Universe do their recycle.

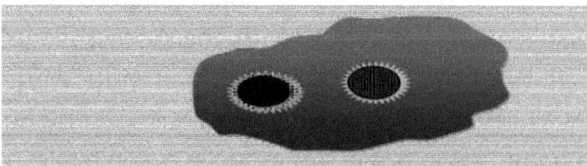

After the Supernova, two new naked black holes split from the
original one black hole at the center of the sun, and each planet
and moon which is obliterated. This doubles the black hole
population with each new 15 billion year cycle

Summary of phases:

Obviously, all of the spheres in our cosmos aren't at the same phase in their life cycle. Each individualistic sphere has its own time frame depending upon many factors. Thus, an entire huge galaxy sees the recycling phases of its solar systems and spheres, at differing paces - not all at one time.

Countless stars and their solar systems recycle on their own cue at 10 to 15 billion years. Most often these events occur far from other stars and solar systems, so they don't affect another system, although the debris from an exploded planet or a rogue black hole vanquished by the Supernova of a solar system could intrude into an adjacent system.

Imagine this scenario: A calm solar system where the sun is shining and all its planets and moons are slowly accepting light from their sun. All the moons and planets are in their lock-up stage of their existence except for the sun which is in an escape phase. Now enters a rogue black hole, flung there from some adjacent solar system where a Supernova occurred. This entering black hole is now caught by the sun's gravitational pull, which holds the black hole in an orbit. This black hole is in a nakedly hungry mode. It attacks the sun of the solar system by madly devouring the sun's light, causing an early Supernova of the sun.

This is a strange occurrence, because the sun holds the black hole (its killer) in its orbit, preventing it from leaving, while the black hole devours the sun.

A long stream of light is seen leaving the sun and then entering the naked black hole. It is a beautiful sight, this stream of light, so bright and lit-up by the massive amounts of light gushing into the bowels of the naked black hole.

For the naked black hole, this is a bonanza, a quick fill-up. For the sun going Supernova from the attack, it is a quick death, dramatically lessening its expectant lifetime cycle.

So much light leaves the sun in a rush that the planets also experience over-heating. Planets and moons closest to the sun are easily destroyed. Middle distanced spheres are hastened to an explosion point as their unraveling of atoms speeds up as well. Only the very most distant sphere(s) in that solar system might survive or be flung away into space. Eventually, the entire solar system, especially those planets and moons closer to the sun, are ablaze, set onto their own path of heading towards Supernova.

As the spheres in that solar system experience their own destruction, the black holes at the core of every sphere are freed of all matter. They become naked black holes once again; ready to devour light inside the remaining graveyard.

Which black hole will get the upper hand by devouring the most light and become the central sphere of a new solar system. This is survival of the fittest to become the master (Queen Bee) of the new solar system.

Interviewer: "Let's say that TT is right in placing a black hole at the core of every cosmic spheroid. Why have astronomers gotten their theories so wrong?"

Author: "Even the very smartest minds can sometimes get stuck with a paradigm such as Big Bang. Many top people totally accept Big Bang, so much so that every new shred of new evidence has to somehow fit the mold. These smartest minds have been conned by these supposed proofs. But, TT argues that those Big Bang proofs have been misinterpreted. TT suggests, keep digging deeper."

CHAPTER 8
SOLAR SYSTEMS
VIA BLACK HOLES

A typical Solar System with sun, planets, and moons.

I n our cosmic trillion year history, black holes first spun atoms to build individual spheres; then the black holes at the cores of these spheres grouped around suns to form solar systems; then these solar systems grouped around central supermassive black holes to form colossal galaxies. This was the progression followed by black holes during the building; atom world (governed by light spinning inside an atom); to spheres (governed by a black hole at the sphere's core); to solar systems (governed by the x-large black hole at the core of a sun); to galaxies (governed by supermassive black hole at the bulging central hub).

This chapter goes back to the early days of our cosmos, to see how the very first solar systems were formed by black holes. There is a most interesting battle and cooperative that goes on between black holes in building a solar system.

Within all solar systems, there exist many small to large spheres. Why this size differential? Why aren't they all just the same size? The answer is that evolution has shown how the fittest grow stronger and larger, and evolutionary growth is prominent throughout our universe, even in the world of spheres and black holes. For example: The supermassive powerful black hole at the central core of a spiral galaxy evolved that way over hundreds of billions of years as it used its more dominant size to overwhelm a multitude of lesser black holes in forming an entire galaxy.

So, "Why then isn't our cosmos just one gigantic solar system or just one super galaxy holding all of the spheres?" Why are there billions of solar systems and galaxies? To answer such questions, we first need to know the origins of the solar systems of our cosmos.

So, we put forth a very intriguing preliminary question to make us think of solar system from a strategic point of view; is our solar system a cooperative or a battle scene?

Most people, after some thought, would normally answer, "A **cooperative,** because the sun, planets and moons in our solar system appear to exist harmoniously, working together in numerous ways. There appears to be no fighting going on between any of the spheres."

TT answers back, "Yes, correct for right now, as our solar system is presently in a cooperative phase between all the spheres during this phase in the 15 billion cycle of our solar system. However, that wasn't the case when the previous solar system ended and the battle commenced by all the naked black holes to rebuild into our present solar system."

It is during this **battle** scene between naked black holes where the structure of all new solar systems is determined. A black hole competes against other naked black holes to win the battle of pulling in energy and converting it into matter.

The battle scene.

Warlike is a way to describe the battle scene phase where a solar system is being formed. Naked black holes battle against one another to claim, devour and spin the most light into matter to build their new bodies.

This battle phase occurs right after the death of a solar system with the definite purpose of building a brand new solar system to replace the vanished one. A short phase, this battle scene lasts mere decades. A gargantuan battle occurs between the naked black holes which were cloaked at the core of the sun and every planet/moon of the old system pre Supernova of its sun and Obliteration of all the spheres.

Now, these naked black holes, unburdened by matter on their surfaces, regain their fast powerful spin as they struggle against nearby opponent black holes for their strongest position in the next new solar system. This battle will rage until the entire structure of the new solar system has been settled and established and all the black holes are cloaked inside of the cores of the newly built spheres. Then, all the filled-up spheres enter the subdued cooperative phase of the next 10-15 billion year cycle of the new solar system.

The forming structure of this new solar system follows the cosmic rule of 'size matters' as the largest naked black holes win the dominancy battle for light and thereby form larger bodies around themselves. These dominators possess the strongest most far-reaching gravitational pull. The largest strongest black hole (sphere) forms the central sun and the next largest black holes form planets, while smaller spheres become moons. In Cosmic Laws, gravity dominance prevails as no moon is allowed to outsize the planet it orbits.

The cooperative scene.

After the battle between black holes ends, forming the new solar system, a **cooperative** exists between spheres in a peaceful phase when each sphere in the solar system is full.

Cooperative is the much longer phase than the preceding battle. In the cooperative, the built mature moons, planets, and sun(s) of a solar system co-exist in harmony. The order of rank within the cooperative is determined by gravitational power of the black holes at the core of each sphere, As well, proximity distance between spheres plays a secondary role in determining a spheres position within a new solar system.

In our solar system, the various gravity tugging going on between all the spheres is presently a peaceful cooperative. As expected, the largest spheres possess the most mass and the strongest gravity allowing them to hold lesser sized spheres in orbit. Our entire solar system appears cooperative as 8 planets orbit our sun's gravity and each planet holds moons in orbit depending upon a planet's gravity strength.

Proximity of gravity plays a role in determining a spheres position within our solar system. Any planet can use the proximity of its own gravity thereby trumping the gravity of the sun in order to hold moons in orbit around the planet. Over 160 moons are now catalogued in our solar system.

Our cosmos is both a fight between naked black holes and then a cooperative between spheres. The battles serve to establish organizational structure of solar systems. All this battling, followed by a cooperative, is governed by Cosmic Laws setting the rules of engagement: rules for the battle between naked black holes in building a solar system; and thereafter rules between spheres in operating a solar system.

First tiny solar system began the first 15 billion year cycle a trillion years ago.

TT shows the Cosmic Laws which the first solar system obeyed. A trillion years ago, an ocean of light limitless in strands in every possible known direction occupied the entire known physical cosmos. This light ocean was frozen full of static light. At this point, there weren't any such things as spheres or space as yet, only the endless ocean of light. (Recently, scientists brought light to a standstill in the lab thereby demonstrating that static light is possible).

An ocean of frozen static (non-moving) light was all that existed at the origin of our cosmos a trillion years ago.

Right thereafter, the first black hole appeared in the ocean of light, providing the prime historical moment when cosmic action really began. That initial black hole possessed an axis which spun at tremendous velocity. As the engine of the new cosmos, that first naked black hole utilized its powerful spin to loosen frozen static stands from the light ocean. Once freed, those light stands overcame inertia and then quickly accelerated. But, before they could make their escape, the rays were susceptible to immediate pull from the naked black hole. Captured, the light strands were spun into atoms of matter deep into the bowels of the black hole. More light was gobbled up as the black hole formed a body of matter.

The constructing CREATION phase had started with the building of the first cosmic sphere. Eventually, the black hole completed its spinning of light and moved to its extra long LOCKUP phase where it would hold matter in place.

Space had been created around the orb as an empty byproduct left over from the disappearance of millions of tons of light from the static ocean of light. In simple terms, the black hole had eaten a cavern of space inside of the non-moving ocean of light. The devoured light, spinning tightly as matter around the sphere, now occupied less than a trillionth the area of the static light which it had devoured.

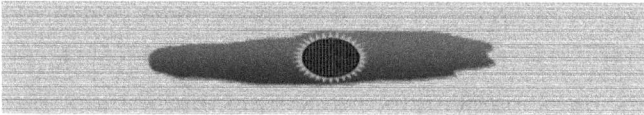

The first black hole was introduced into the static ocean of light. It loosened static light from the ocean, freeing that light and spinning it into matter around itself, leaving empty cold dark space in the area close around itself. As that first black hole continued to devour light, it grew into universe's first planet.

Next, a second black hole was introduced into the static ocean of light, near the first black hole. The second younger black hole revolved around the first black hole which had grown into a planet. That planet's strong gravitational efforts to hold its contents in place extended outward to keep the young black hole in orbit. The second black hole attracted and spun light into matter from the static ocean of light, but it also hastened the loosening and expanding of the planet.

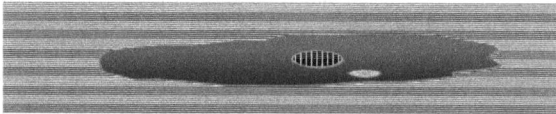

A second black hole was introduced, orbiting around the already formed planet (larger sphere). First tiny solar system formed.

The small naked black hole ate away at more available light from the ocean of light, but its ultra-fast spin also pre-loosened the atoms on the planet.

The planet's loosening atoms hastened its transition from a hard-surfaced planet to an expanded liquid surface. The planet lost more control of its atoms with a steady stream of light leaving its surface to be devoured by the new black hole. The first sun of our cosmos was the result.

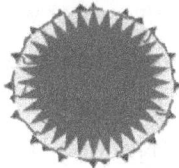

The planet's transformation to the star stage was hastened.

Eventually, the sun expanded immensely while it still held the eating black hole in orbit. Therein, the first solar system was born with one central parent sun (with the original black hole inside its core) and one small orbiting childlike planet (with the other black hole inside its core).

Over the coming eons, the new small planet devoured more light from the sun further loosening and expanding the aging sun. Ultimately, the sun over-expanded and went Supernova and Obliterated the planet which it held in orbit. As the process completed, the sun rid itself of its matter which unspun back to light. The orbiting planet was burned to a crisp (obliterated) by the exploding sun such that all of the planet's atoms of matter exploded and went back to light. The first solar system came to an abrupt ending.

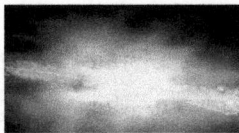

First the sun went Supernova and exploded.
The Supernova effect melted the planet into Obliteration.

This may seem strange for the star to extend gravity out to the young black hole to keep in orbit. All the while, the young black hole freely preyed on and devoured the star. But in the end that, turnabout was fair play. The star said, "Yes, you can eat me, but I will hold you in orbit, and when I'm forced to explode, your matter will be annihilated too."

On the heels of the Supernova and Obliteration, there was no evidence left around as proof showing that the first solar system had ever existed. No long-term graveyard or markers recording its history. The recycle left no historical fossils.

Except that is, for the two surviving black holes: one had been at the core of the sun; the other cloaked inside of the planet. The sun's black hole had survived going Supernova and the planet's black hole survived Obliteration.

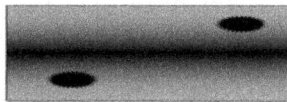

Two naked black holes survived Supernova
and Obliteration of our universe's first solar system.

The second solar system (2nd 15 billion year cycle).

A black hole can never be destroyed. Thus, the two black holes of the first solar system survived a Supernova. These two inaugural black holes each then subdivided.

The four new naked black holes of the region immediately commenced a battle for light from the nearby static ocean of light and also from light attempting an escape from the sun's Supernova and the planet's Obliteration. Each naked black hole spun light into matter adding to their length and girth of their axis and the size of their cores. Each of the new black holes in this 2nd cycle was larger than in the first cycle.

Of the four new naked black holes, the one being the best positioned and most opportune consumed the most light and was the first to shift its main focus from consumption of light to the long task of holding matter. It also gained the most overall mass of the group holding the other 3 in orbit. It was destined to grow into the sun of that new system.

In the long run, this second solar system faced the same fate which had befallen the first solar system. In solar system two, the three spheres held in orbit by the largest sphere were still in devouring mode. They loosened the atoms of the central sphere which was already in control mode. After billions of years, that central sphere loosened from a large planet to star. After more billions of years it loosened as a star and went Supernova. Likewise, that second solar system also met with Obliteration. The devourers of the central sun caused it to go Supernova and it played the same dirty trick on its devourers as all were obliterated by the Supernova.

Of course, Cosmic Law once again prevailed as black holes at the core of the sun and planets survived. There were four surviving black holes after Obliteration of the 2nd system. Each was already in motion working hard to reproduce by fission and split into two new naked black holes, upping the count of naked black holes from 4 to 8.

The third Solar System (3rd 15 billion year cycle).

In the third solar system, cycle three, there were 8 new naked black holes. The sphere population had doubled once again, ready to form yet again a new larger solar system with one central sphere and seven spheres in orbit. With 8 spheres in play, the Universe Laws which existed between black holes, and planets, and suns, were put to a sterner test.

As usual, the black hole which was the largest dominant eater of light became the focus of the next solar system. Over the next few billions of years it grew enormously from being the largest planet in that new system to being the sun.

Other changes occurred as well: The ocean of light was now further from the grasp of any of the black holes which were growing their bodies into planets. Thus, the new solar system had a more spread out appearance. While the central largest planet (sun) held the other seven black holes in orbit, the black holes were pulled outwards by the attraction of the light ocean on the perimeter. Expansion of the cosmos was in full swing. More light was consumed by each larger solar system which stretched itself further from the center of the cosmos to stay close to the perimeter's ocean of light.

A secondary dominant player appeared. While the largest planet became the sun of the new larger solar system, that sun had difficulty controlling each and every one of the planets in orbit in its system. One larger planet exerted its powerful gravitational pull on a smaller sphere forcing it into orbit as a moon. So, this 3rd solar system had the structure: 1 central sun; 6 planets; 1 moon = 8 spheres. At the end of cycle 3, Supernova and Obliteration were more explosively super because this was a larger sun than in earlier cycles.

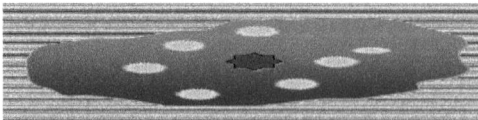

Solar System 3 (8 spheres). The static unmoving ocean of light on the perimeter surrounds empty space which the spheres had created by consolidating light into matter around them. One central sun, six orbiting planets, plus one small planetary moon, comprise the 8 spheres in solar system 3.

The fourth 15 billion year cycle.

In cycle four, the count of black holes doubled again, this time from 8 to 16. This time the black holes had been blown even further apart, leaving various possible formations.

Scenario one: one larger system comprised of 16 spheres.

Scenario two: The two largest survivors from the old sun going Supernova are the two contenders vying to be 'Queen Bee' as the central force(s) of the new solar system. Two spheres control the solar system becoming its binary suns.

Scenario three: The two largest surviving twin black holes from the Supernova are driven far apart by the explosion. The lesser dominant black holes follow either of the two leaders, forming two new solar systems. If one system is larger than the other, the larger system extends its greater gravitational pull to hold the smaller system in a distant orbit. As a direct consequence, the first cosmic galaxy forms.

Some Solar System Rules (Cosmic Rules):

• Holding gravity of a sun to keep planets/moons in orbit depends directly upon its mass. Gravity G, the gravitational constant of the universe, times the mass of the orb, divided by the square of its radius. Gravity = $G(M/r2)$.

• Pimple Burst always occurs on planets and moons. Spheres can only maintain their holding of matter phase for a certain period. Eventually, the loss of control of matter begins deep in the sphere's core. Melted matter (lava) searches for an exit. As eons pass, pressure intensifies; the sphere can either explode or depressurize. Easing comes via the initial Pimple Burst volcanic eruption. With titanic force, the daddy of all volcanoes propels massive amounts of debris into space to pelt down and crater onto other spheres of the solar system.

• The distance between spheres becomes greater with each recycle into a new solar system, or systems.

• Black holes of new solar systems always grow larger than preceding black holes of past systems.

Each of the subsequent 15 billion year cycles doubled the size and population of the cosmos.

As each of the 15 billion year cycles of the universe took place, each brand new cycle saw the sphere population doubled and space widened within each new solar system or in larger venues called galaxies.

But, not all of a galaxy recycles at any one time. Galaxies are so enormous that their solar systems are totally remote from one another. Approximately every 15 billion years, all zones of the RECYCLIUN universe recycle, thereby doubling the sphere population in that zone, thereby contributing an ever increase to the size and population of our cosmos.

****

HOW OUR SOLAR SYSTEM FORMED

In our solar system, we are in a long peaceful **cooperative**. But, a few billion years hence, when our sun goes Supernova, explodes and destroys our solar system, the **battle** between all the surviving naked black holes will start with the goal for every black hole to rebuild a new sphere around itself.

This battle will see a whole new restructuring of the next new solar system. The battle may even see a powerful split with two new solar systems replacing our present one.

In this trillionth birthday of our cosmos, our solar system is one of millions in our mammoth Milky Way. Trillion Theory states that we currently reside in the 67th of the 15 billion year cycles which took place over a long cosmic history.

Now, we revisit the start of our present solar system, 10 billion years ago. The previous 66th cycle had just ended.

So, when our new solar system was replacing the old solar system which for 15 billion years in cycle 66 used to occupy this area of our Milky Way Galaxy, a percentage of other solar systems within our galaxy were doing likewise. Therein, the overall population of Milky Way was growing.

When the old previous solar system in this locale aged out, the sun at its center went Supernova. Cosmic Law was instrumental in determining how that Supernova occurred, the effects it had, and what happened right after. Cosmic Law also controlled how our new solar system would form.

Events occurred for the old solar system:

Step 1: The old sun shed all of its matter at one Supernova moment, releasing it back to straight line light.

Step 2: When the old sun went Supernova, the cloaked black hole which had been at its core survived and split into two new naked black holes. These two naked black holes were driven far part from one another. Both immediately began a search for light to spin and build into new spheres.

Step 3: All other spheres (planets/moons) in the old system were destroyed by the Supernova and Obliteration, as they too shed matter back to light. They all went Supernova as well. Over 100 new naked black holes were the final result. That Supernova, followed quickly by that Obliteration would mark the end for that old solar system at the end of cycle 66.

Steps occurred in forming our present new solar system in the 67th 15 billion year cycle of our physical cosmos:

Now that the old solar system at the end of cycle 66 was gone, we follow how our solar system formed at the start of cycle 67. Cosmic Law will guide us by showing why we have a central sun; why planets and moons are located where they are; why our solar system is quite linear in shape; why all the planets rotate in the same direction around our sun; why the different spheres have different sizes, axial tilts, and surfaces; and why planets rotate on their axis. Basically how our whole solar system developed to its present situation.

Our Sun, an x-large black hole formed into our Sun.

The x-large black hole which was destined to become our sun (which was also at the center of the old sun) was the first ready naked black hole after the Supernova to devour light. The other black holes at the center of the old planets/moons all faced Obliteration. Their matter would unravel back to light only to be quickly devoured by the x-large black hole.

That x-large black hole was the most gluttonous greedy voracious eater of light. It became the Henry VIII of our solar system. Basically eating until it was ready to bust. It hogged light, using its size, early positioning, and gravitational pull to steal light from the grasp of others.

That x-large black hole stored the captured light as new matter. This light entered the x-large black hole's bowels as 'coming in hot' (light carries the heat property). So much light was being consumed by the x-large naked black hole in a short period of time (Quasar) that most of the atoms were 'staying hot.' The earliest 'hot' atoms were layered over and kept hot by the next blanket of atoms spun over them.

Our sun is much larger than any of the planets because: it has the largest black hole at its core; that black hole spun tons of light in short time which it captured from the nearby source of the dissolving planets and moons; that light 'came in hot' in abundance and stayed hot making for the loosest spun atoms (occupying more area) of any of the spheres in our system. (Cosmic Law: the larger a black hole the looser it spins its atoms, layer by layer, from inside to outside).

As the largest black hole of our solar system, our sun gained the most mass and the central position as it held the other smaller black holes in orbit. But, as that black hole gained mass and it grew into a sun, its mass empowered it to hold all the forming planets in more permanent orbits.

While the x-large sphere utilized its powerful gravity to hold the other spheres in orbit, the other smaller black holes were still in the eating guzzling stage and they attacked the loosened up x-large sphere which was holding them in orbit. Their attack further loosened the atoms of the x-large sphere turning it into a sun to unspin and emit light.

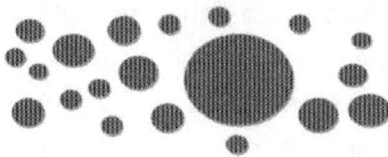

When a star goes Supernova and its Obliteration destroys the old solar system, history is wiped out as no graveyard remains. All that survives are naked black holes, poised to rebuild a brand new solar system. At this stage, the black holes are in a scattered pattern. Axial tilts can vary as each black hole best positions itself to attract light from the Obliteration, or from the static ocean of light if it is nearby, or emitted from distant stars.

Interaction between planets and their sun.

Planets in our solar system play second fiddle to our sun. Our sun's cloaked x-large black hole out witted all the lesser sized black holes inside all planets and moons when it came to attracting and spinning light into matter to form a sphere.

These smaller naked black holes entered into battle with our sun, which had become so full that it had moved from spinning light to lock-up while trying to hold its contents.

The naked smaller black holes were spinning fast, looking to capture light, and the gravitational pull of their spin was further loosening atoms on our sun's surface.

The smaller naked black holes pulled much light from the sun, spinning that light into matter. They would have forced our sun to go Supernova right then and there except for one thing: as they became fuller forming into planets/moons, their spinning of light into matter slowed to a trickle.

Thus, our solar system was saved and moved to what it is now a sun steadily emitting light and planets and moons accepting that light only to warm their surfaces.

Our sun spun so much light into matter so quickly and that matter 'came in hot' in such large volumes that our sun forfeited a chance for matter to cool. In contrast, moons and planets spun their matter slowly. Each new layer had more time to feel the coldness of space before more hot was spun overtop of the previous layer. Therein, only the center of moons and planets has remained hot.

However, even for planets/moons that internal hot always expands as more atoms loosen on the interior. Loosening of interior atoms always wins out over the coldness of space. Spheres are never solid cold, a hot core always percolates.

How our Solar System came to be organized.

The organization of our unique solar system followed a set of quite common Cosmic Laws. Our solar system has a sun with 8 planets (listed below from closest to furthest from the sun; and with diameters and number of moons):

Mercury (3,032 miles in diameter)(no moons)

Venus (7,521 miles in diameter)(no moons)

Earth (7,926 miles in diameter)(1 moon)

Mars (4,222 miles in diameter)(2 moons)

Jupiter (88,846 miles in diameter)(63 moons)

Saturn (74,898 miles in diameter)(62 moons)

Uranus (31,763 miles in diameter)(27 moons)

Neptune (29,700 miles in diameter)(13 moons)

Total of 177 spheroids; 1 sun, 8 planets, 168 moons. Why are the spheroids where they are? Accidental? No. Cosmic Laws tell us that smaller spheres are more easily pulled close to our sun. Whereas, larger spheres stay further away.

Organization followed Cosmic Laws pertaining to black hole sphere size and gravity pull. Mercury (smallest) was the sphere most easily pulled close by the sun's gravity. Mercury was too small to hold orbiting moons. Venus, also no moons as it was overpowered by gravity from the close in proximity sun; Earth 1 moon and Mars 2 moons are a little further away from our sun's powerful gravity. Then, the gas giants were large with enough mass to hold many moons and challenge the distant sun's gravity; Jupiter has 63 moons; Saturn 62; Uranus 27; and Neptune 13. Saturn is the most interesting planet with its rings circling around its equator, illustrating cosmic gravity laws around a sphere.

Our solar system displays a rather flat pinwheel shape.

Our sun is the center pin and the planets orbit at varying distances from the sun. These planets are in a flat plane because of TT Cosmic Law stating that black holes exert their strongest gravity pull extended out along their equator.

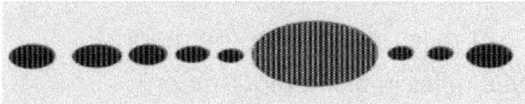

Our formed solar system took on a linear shape because of the black hole at the central core of our sun. The spin of that black hole rotates the sun, but also creates a gravitational pull holding the planets in orbit extending out along the sun's equator.

Cosmic Laws which spheres in our solar system obey:

After a naked black hole fills from eating light, its gravity duties change from spinning light to the holding of those manufactured atoms in place. This gravitational pull of the black hole at the sphere's core extends out past the surface in a linear flat plane extending out around the equator. A larger sphere will have greater mass and be able to extend its gravity further and be able to hold more spheres in orbit.

The x-large black hole at our sun's center possessed the strongest gravity to win the battle to attract light, spin it into matter, and build a massive body around itself. Then, it began exerting its strongest gravity force onto the other lesser black holes pulling them into orbits around itself.

The orbital direction for the 8 planets of our solar system followed Cosmic Law stating that all planets held in a sun's gravity revolve in a similar direction determined by the direction of spin of that sun. Our sun spins counterclockwise on its axis, thus all the planets orbit it counterclockwise.

However, moons are first and foremost tied in orbit to the axial direction of spin of the planet which they orbit. That orbit could be clockwise or counterclockwise relative to the planet depending upon the planet's direction of axial spin.

Note: Here, it is important to differentiate that spin is different than orbiting. A main motion that a sphere has is the spin pivot rotation on its own axis. The direction and speed of that rotation is caused by the direction and speed of rotation of the black hole at the sphere's core. Thereafter, the gravity of other spheres can effect it, such that a sphere can revolve (orbit) around a larger sphere. The direction of orbit is determined by direction of spin of the larger sphere. Venus spins clockwise on its axis since its black hole spins clockwise. But, Venus revolves counterclockwise around our sun because of the sun's counterclockwise axial spin.

Here, TT objects to both Big Bang and Nebular Theory. If a circling nebula cloud caused our solar system, then all the planets should have been imparted with a similar direction of axial spin. Since our sun has counterclockwise axial spin, then every planet should have counterclockwise axial spin.

But this isn't so. All the planets and moons in our solar system don't spin the same direction on their axis. This is so because our solar system didn't form from a swirling nebular cloud. Trillion Theory shows that spin direction for a sphere is totally dependent upon the direction of rotation of the black hole at its core. This direction is determined when a naked black hole begins a new cycle. In our solar system, the black holes at the cores of Mercury, Earth, Mars, Jupiter, Saturn, Neptune, and our sun rotate counterclockwise on their axis. But, the black holes of Venus and Uranus don't.

Weirdo bizarre Venus and Uranus spin CLOCKWISE. They make a nightmarish unexplained mystery for astronomers. They cast substantial doubt on Big Bang and Nebular theory.

That mystery is best explained in Trillion Theory. Both Venus and Uranus, the so called weirdo planets, simply have black holes in their cores which spin clockwise. (Note: The direction of spin of spiral galaxies has been found to be 50-50, indicating that half the supermassive black holes at the hub of galaxies spin counterclockwise, half clockwise).

Now, Cosmic Laws within TT even deal with the speed of spin of spheres on their axis. Of course, this is tricky since the speed of spin of a sphere on its axis changes over time. Naked black holes, small or large, possess the propensity to spin at the highest speed. But, once full, the contents which they control can slow their rotation, as can the influence of other powerful black holes in the near vicinity.

In writing 'Rotational Speed Laws,' Trillion Theory keeps them separate for naked black holes and spheres with mass:

Rotational Speed Laws for Black Holes:
• Naked black holes have a super-fast rotating speed.
• Fuller black holes witness a slowing of their pivotal speed.

Rotational Speed Laws for Spheres:
• After a black hole body builds a body, its spin slows.
• Partially full black hole: sluggish speed.
• Full spheres: even more sluggish speed.
• Suns are more sluggish than planets.
• Larger suns are more sluggish than smaller suns.

Interviewer: "It would seem easy as pie to define every solar system. I'd say a solar system is an accumulation of planets and moons orbiting around a central sun. Correct?"

Author: "Almost. Your definition is acceptable for an evolved solar system such as the one our Earth is in. However, not all solar systems are in this same exact phase of their cycle. Our solar system is now in the 5-10 billion years of its probable 13-15 billion year cycle. However, an accumulation of black holes fighting over light and trying to figure out who will be the 'head honcho' to have all the others orbiting around it, is also a solar system, just in the earliest cycle stage."

Interviewer: "So, is there a limit to a solar system's size?"

Author: "Yes, theoretically so. The powerful split-force when a large solar system is Obliterated by a Supernova separates the old solar system into two new ones. This is evidenced all across our cosmos by billions of solar systems. Also, galaxies initially result from the splitting of overly large solar systems, with the larger of the two holding the smaller one in orbit."

Interviewer: "I'm following your explanations, but your view of how the first-ever solar system grew from the static ocean of light a trillion tears ago seems rather far-fetched to me?"

Author: "Any more far-fetched than the Big Bang? Please be patient, wait for my entire theory start to finish, then pass judgment. Remember, TT proclaims that the design of our cosmos was done with simplicity as well as strategy. If we can prove that light and black holes do what TT says they were programmed to do, that will prove Trillion Theory."

Interviewer: "I was excited recently to hear of the existence of a multitude of solar systems even in our Milky Way. Do you think that TT increases the potential for other life?"

Author: "Most definitely so. In our cosmic trillion years, life has abounded. TT's supposition that there has been many recycles of our cosmos raises that likelihood beyond belief."

CHAPTER 9
GALAXIES BUILT
BY BLACK HOLES

Spiral galaxies are a near perfect organized island colony.
Trillion Theory introduces a new understanding of galaxies.

Galaxies are definitely the most ominous beautiful sights in our cosmos. They present themselves with power, magnitude, majesty, and magnificence as they show off their splendor and flaunt the beauty of their spectacular array. Galaxies are wondrous yet solitary. They can be described as gigantic remote private islands of the cosmos. Our present cosmos is estimated to hold two hundred billion galaxies. One galaxy may contain millions of stars with solar systems.

While solar systems recycle on average every 15 billion years, galaxies are much more permanent features because they don't live and die by the 15 billion year rule. It is solar systems which recycle within the galaxies of our cosmos each 15 billion year cycle thereby fostering growth. Galaxies remain more permanent features and grow in size each time one of their solar systems recycles. Continually, solar systems become ever more populous throughout their galaxy.

So, think of a galaxy as one of the many island hotels of the cosmos. A galaxy may have millions of hotel guests residing as solar systems. No matter how many times the guest occupants (solar systems) of the hotel rooms change, that remote galaxy hotel remains intact as the main entity.

Once a galaxy becomes gigantic, its configuration easily survives a Supernova recycle of one of its solar systems while the rest of the galaxy remains intact and unaffected. While various parts of a galaxy recycle every 15 billion years, the galaxy itself may be hundreds of billions years old.

The oldest largest galaxies have gone through as many as 30 to 50 of the 15 billion year recycles of their contents, over the past 400-800 billion years. Estimate the number of stars in a galaxy and it is possible to closely calculate the age of that galaxy using the doubling premise each 15 billion years. Galaxies are the best recorders of cosmic history. They have witnessed vast growth over hundreds of billions of years.

Spirals make up about 90% of our universe's galaxies. Their stars and solar systems are in orbit whirling around a galactic bulging center. This bulge has enormous mass with enough gravitational power to hold in orbit millions of stars and solar systems. At the core of the bulge is a fast spinning supermassive black hole middling a large concentration of densely clustered superstars. That bugle is exactly the spot where Cosmic Laws pertaining to black holes are rewritten.

But, don't think of a galaxy as simply a huge solar system. A solar system has spheres orbiting a sun; whereas, a galaxy has many solar systems orbiting a supermassive black hole.

Also, in a galaxy, the supermassive black hole centering the galaxy has evolved to a definite new level of existence.

A supermassive black hole centers a galaxy. It is the monarch of the galaxy where it builds an empire of solar systems.

This chapter will show the evolution of a black hole which becomes the monstrous sized central bulge at the core of a galaxy. Cosmic Laws are re-written by the supermassive.

History building up to galaxies:

A trillion years ago our cosmos started from one small solar system growing into more, and eventually into billions of galaxies. We detail key 15 billion year recycle points:

• The first solar system had 2 spheres (sun and planet).
• That first solar system grew in size entering cycle two.
• The second solar system grew in size entering cycle three.
• The third solar system grew in size entering cycle four.
• The 4th solar system, split into 2 solar systems for cycle 5.
• The largest sun kept all the other solar systems in orbit.
• That formed the first small galaxy of our cosmos.
• At the center of that small galaxy was a massive sun.
• At the core of the sun was a massive black hole.
• The massive went Supernova and Obliterated the galaxy.
• Two new galaxies formed replacing the old galaxy.
• Black holes at cores of galaxies became more massive.
• Massive black holes evolved to being supermassive.
• Super size brought structural changes to the black hole.
• These supermassives evolved new laws for themselves.
• These supermassives learned how not to go Supernova.
• These supermassives learned how to become ageless.

Evolution from solar system to galaxy in Trillion Theory.

TT has shown how our cosmos began small a trillion years in the past. Naked black holes ate from the static ocean of light to form the first solar system. This system grew with each recycle into several more solar systems. The first galaxy formed when several solar systems came under the control of the central massive star of the largest solar system. Then, the earliest small galaxy split apart forming two galaxies. Today, there are billions of galaxies housing solar systems.

During this progression in numbers, the main change which occurred was the size of the most dominant sun of the largest solar system. Every 15 billion years the massive sun centering the largest solar system grew ever larger.

However, the true progression which occurred was that it was the black hole at the centre of the massive sun which grew proportionally more dominant each cycle capable of forming a very massive sun. And, that black hole graduated to a supermassive size, capable of controlling a galaxy.

At a later point in its evolution, this supermassive black hole determined not to eat light solely for the building of a supermassive sun around itself – that was too unproductive. It found that building a sun meant eventually having to go Supernova and detrimentally destroying its galaxy. A smarter strategy presented itself, allowing supermassive to evolve much older and control its galaxy for longer than just one 15 billion year cycle. Instead, supermassive's galaxy could grow for 100's of billions of years. Supermassive, to complete this feat needed to change its strategy for conducting business.

Perhaps supermassive's programming could change the rules it played by when achieving a certain benchmark size.

Galactic Obliteration.

What is Galactic Obliteration? Why is it relevant?

When the galaxy is still small, the humongous star that centers the galaxy can go Supernova and make every star and solar system within its galaxy face Galactic Obliteration, death and rebirth of the entire small galaxy. This is by far the brightest event ever as an entire galaxy goes ballistic with a multitude of Supernovae exploding simultaneously.

In certain ways a Supernova creating Galactic Obliteration was good. It was a way for an entire small galaxy to instantly double its number of solar systems.

The bad was that after the Supernova by the massive sun centering the old galaxy occurred, too often that powerful Supernova split the galaxy into two, pushing apart the two largest massive black holes. Each of the two massive black holes took half the lesser black holes along to form a new galaxy. Thereby, two new galaxies receding away from each other replaced the old galaxy. Thus, it was difficult for any one galaxy to remain intact and grow ever larger.

Also bad was that after each gigantic Obliteration a battle raged by all of the massive black holes during Galactic Reformation to see which one would become the master (Queen Bee) of the new galaxy. With Queen Bee changing every 15 billion years, any newly crowned Queen Bee had to discover a method to remain on the throne as long as possible as a supermassive black hole centering the galaxy.

The good was that sometimes after Galactic Obliteration, the galaxy stayed as one and grew ever larger. Its remote solar systems went Supernova at separate times without affecting the central dominant massive black hole.

Each new Galactic Reformation saw the central black hole more disproportionally larger in ratio compared to all its other diminutive rival black holes. Eventually, the prominent master black hole (Queen Bee) had hundreds, to thousands to millions of solar systems in its galaxy. So strategy-wise, destroying all at a moment's notice was not productive for the Queen Bee supermassive black hole. That wasn't sound strategy, and black holes are all about strategy. And the larger a black hole becomes the more cards it holds and the harder it can trump any other smaller black holes.

Finally, the point came for the supermassive black hole (Queen Bee) where it wanted to maintain that which it had built. The definitive strategy for the supermassive black hole was to adjust its structure giving it new powers. These new powers allowed the Queen Bee supermassive black hole to overcome the rule which had always prior taken it into the recycling process every 15 billion years.

Breaking the rule of 15 Billion Years. The supermassive black hole (Queen Bee) of a large spiral galaxy changed the rules allowing itself to grow far older.

All moons, planets, suns and solar systems recycle every 15 billion years in the Recycliun Universe, and initially so did the massive suns centering galaxies. But, for any sun to avoid going supernova, the massive black hole at its core had to develop new strategies. The cornerstone of this new strategy called for the massive black hole to be able to control the galaxy without having to grow into a massive sun itself, and then later have to go Supernova. Note: It is suns which every 15 billion years weaken and go Supernova. Whereas, the XL black hole at the sun's core always survives.

So, the soundest strategy for a massive black hole at the center of a galaxy was to avoid having to grow into a sun. Step one in this new strategy of breaking Cosmic Law was to do something totally different with the tons of light which the massive supermassive black hole pulled inwards from other suns. This required a structural change for the massive black hole by evolving structural transformations.

A quick review: Only a small percentage of black holes ever evolve enough to make it to massive status. First, a black hole must become dominant as a sun in its solar system. Then, as a next step become the dominant XXL black hole at the core of a sun in control groups of solar systems in a small galaxy. With each step, the black hole grows in size dominance. Its central helix becomes longer and its body core increases in size and in number of compartments.

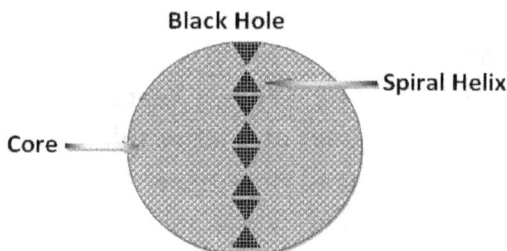

Review structure of a black hole. A central spiral helix at the core of the black hole pumps up and down while it rotates a core around itself made up of trillions of compartments.

Thus, a supermassive black hole which didn't desire to grow into a sun had to develop new skills. The supermassive had to make (evolve) new structural changes. Till then, every cycle the black hole had always added to the girth (breadth) and length of the central spiral helix axis operating its core.

As the black hole evolved further, every part of its spiral helix expanded in length and breadth making the central helix a totally wider more hollow structure on the inside of its form. That helix took on so much extra breadth that a vacant area running the length of the helix developed, similar to a hollow pipe. Eventually, this looseness in the helix prevented it from spinning as much light into matter. The helix finally widened so much that it evolved into more of a conduit where light can pass right through. Basically, now the supermassive black hole only spun light into matter to continually build its internal structures, while never again adding more matter to build a bodily sun around itself.

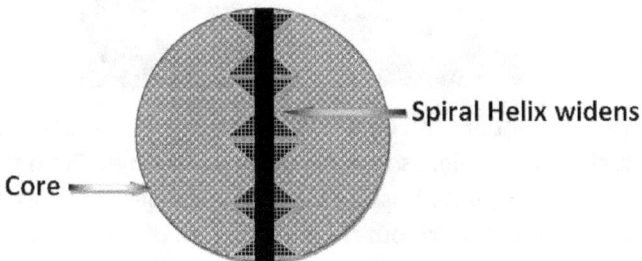

Massive (Supermassive) black hole with a wide hollow expanded area along its spiral helix.

With the development this skill, the supermassive black hole was able to spend more time and focus on continually growing its internal structures, thereby totally moving faster than any other competitive black holes, enroute to growing evermore supermassive. This new strategy also allowed the supermassive black hole to pull in extra tons of light, using only some of this light for building internal structures, while passing the bulk of the pulled in light right through its hollow helix and then spewing it out through its poles.

In essence, the supermassive black hole at the galaxy hub had learned to breathe, attracting tons of light into itself and then able to plume much of that light outwards via its poles.

As the supermassive black hole learns to breathe, spew out and jettison light in plumes from its poles, it never has to grow a huge sun body and eventually have to go Supernova. It has strategized a method whereby it can live a far longer cycle than just 15 billion years and maintain its dominant position as a supermassive at the galactic centre of a galaxy.

Plumes of light are forced out the top and bottom of the super-massive black hole, as its helix becomes hollow. The pumping action of the supermassive black hole pushes the loose light from the inner hollow out the top/bottom of the spinning hole. Supermassive has learned to ventilate through its ears.

That was quite the feat for a massive black hole to be able to change its structure and the Cosmic Laws pertaining to it, finally living right through the 15 billion year rule.

A Supermassive centering a galaxy grows much older.

Some supermassive dominant black holes have been at the center of their galaxy for up to 700 billion years. They're likely destined to remain in that dominant galaxy position for ever, never growing into a sun or going Supernova.

The supermassive black hole is keeping down its own fat-like body mass by expelling incoming light through its poles. The changes in function allow for wins on several fronts.

Firstly, the supermassive black hole doesn't have to grow into a sun. It can live long, control its galaxy, and prosper.

Secondly, it can spin fast because it stays as a black hole, never taking on the sluggish rotational slowness of a sun.

Thirdly, it can take huge amounts of light from the galaxy stars, then expel, spew and jettison that light as plumes out through its poles providing fresh light to the gigantic galaxy.

Fourthly, it can use the rotational powers and mass of the stars it pulls close to assist in the spinning of its spiral arms.

So for a supermassive black hole, humongous makes a difference. The rich get richer as the supermassive black hole of a galaxy can live on through all the 15 billion year recycles of the stars in its galaxy. By not having to grow into a sun, the supermassive black hole fends off having to recycle itself and instead it can maintain its stature as the hub of the gigantic galaxy for hundreds of billions of years.

In the next 15 billion cycle of our cosmos (cycle 68), cosmic size will again double. Within known galaxies, solar systems will become larger as spheres are added. More solar systems will be counted as some existing ones will split during their Obliteration forming two new solar systems. The cosmic star count will double from 73 quintillion stars in our present 67th cycle to 146 quintillion stars in the 68th cycle.

At the center of a spiral galaxy, a supermassive black hole bulge holds the galaxy together in a linear pancake formation. This supermassive black hole is surrounded by many large stars. Light plumes out the top and bottom of the black hole.

Clearly seeing a galaxy's supermassive black hole is not always easy. A bright halo surrounds the supermassive.

A supermassive black hole transforms to the point where it can attract more light than ever before. This strategy also allows the supermassive black hole to pull dominant stars into a cluster around itself, eventually unspinning those stars and gobbling up their light. Thereby, never allowing those stars to challenge for supremacy of the galaxy. Although the mass of the stars assist the supermassive black hole with the gravitational pull necessary to hold the galaxy together. In essence, the ingenious strategy of the supermassive black hole uses its prowess to make these stars into servants to assist in the controlling of the humongous galaxy.

These pulled-in stars glow brilliantly like a halo as there is a plethora of stars in concentrated numbers surrounding the supermassive black hole. Note: This halo makes it difficult to get a pure look at the supermassive black hole.

The supermassive black hole's pluming out of light also fits into its strategy. This galaxy might be far from the outer ocean of static light, so the supermassive strategizes closer sources of light making its galaxy grow and ensuring that its solar systems recycle on cue. It speeds this process by pulling large stars in towards itself and sending their light out for resident black holes to use in building solar systems.

However, there is always more and more light needed to run the operation of a galaxy. Thus, galaxies are continually attracted outwards towards the perimeter of our cosmos to be closer to the available ocean of light. That is the scene where many new naked black holes are busy at work breaking off chunks of the static light, trying to consume it.

Some of that uncaptured escaping light from the ocean of light travels through space becoming available to current galaxies. Therefore, galaxies recede outwards as they try to get closer to space's perimeter. Whereas, the inner cosmos is just empty space leftover after light spun into mater.

The billions of galaxies in our cosmos mostly travel outwards expanding away from the center, pursuing the abundance of light available in the outer static ocean of light surrounding the perimeter of created space.

Is perfection within our cosmos ever 100% attainable?

No. Not every piece of debris, whether comets, asteroids, dust clouds, gas or clouds, can be perfectly re-assimilated by naked black holes right after Supernovae or Obliterations occur in solar systems and galaxies. Unaccountable stuff, like the gaseous clouds, might never be assimilated properly. This is the .01% which will be studied for decades after we figure out the more important systematically relevant 99.9%.

Does this make our cosmos too imperfect or too helter skelter? No. But, perfection to 100% recycle is unlikely with untamed debris floating inside of solar systems and galaxies.

The awesome beauty of the recycling cosmos.

Cosmos most often appears like a beautiful still picture. We really see very little fast-action accept for Supernovae; for the cosmos recycles at a snail's pace, and eon at a time.

Interviewer: "At what point in history did solar systems get big enough to be called galaxies."

Author: "It probably occurred about 700 billion years ago in the trillion year history of our cosmos. An estimate would be around cycle 20 (each cycle being 15 billion years) which was 300 billion years into cosmic history. That is the point where there were half a million spheres. These solar systems were likely within decent proximity of one another. That is also the point where the black holes at the center of suns of solar systems would have evolved large enough during their Supernova to blast some of their large solar system far enough away to create a whole new other smaller solar system. Several smaller systems likely orbited around the larger solar system. Hence, the first galaxy."

Interviewer: "It's interesting in TT as to why our cosmos is expanding outwards at an accelerating speed. The Big Bang theory attempts to say that this expansion outwards was from the original Big Bang in the middle of the cosmos. I now see where TT thinks much differently."

Author: "Yes, definitely. TT explains how our cosmos has the galaxies moving outwards continually pursuing the greatest available light from the ocean of light in the outer perimeter. Our cosmos will continue to expand outwards forever."

Interviewer: "How does TT account for the scientific reports of dark matter surrounding galaxies?"

Author: "The supposed identification of dark matter has no decent reason. Astronomers haven't really seen this stuff they call dark matter, yet they're simply supposing that it exists because of readings which they receive when they study the mass and motion of large galaxies."

CHAPTER 10
A NEW EXPLANATION
OF GRAVITY

U nderstanding gravity. Okay, really, what is it? If asked, a person would say that gravity is a force which holds us on our planet's surface and our moon in orbit. All planets are held in orbit by the gravity mass of their sun. This gravity is taken for granted. But, that doesn't explain what causes this gravity to be a cosmic force.

Trillion Theory claims, 'Gravity is the strong gravitational pull which an ultra-fast spinning black hole extends outside of its core when that black hole is consuming light.'

Trillion also says, 'Gravity is the strong gravitational pull which a black hole extends to the surface and out past the surface of the sphere which it inhabits the centre of.'

Which is it? TT says it's both, depending upon the stage of a black hole. TT advocates that, 'Gravity is a property owned by the black holes of our cosmos. Initially, gravity can pull light into a naked black hole. Thereafter, when a black hole occupies the core of a sphere, the gravity which it projects stops things from leaving its surface or its orbits.'

The gravity owned by the black hole at Earth's core originally spun at an ultra-fast rate as a naked black hole when it attracted and spun light into matter to build a body of matter to form Earth. Now, this same black hole existing at Earth's core spins slower since it has been slowed by all its acquired mass. Yet, it still extends its gravitational pull to hold us on the surface and also hold our moon in orbit.

Largest black hole of a solar system controls the system.

The size of a black hole determines the power and reach of its gravity. Gravity follows specific rules: larger black holes have a stronger gravity and can extend their gravity further afield. This means that a larger black hole sphere can trump and hold a smaller nearby black hole sphere in its orbit.

The XL black hole at the core of our massive sun forces all planets in our solar system with smaller black holes at their cores into orbit around it. And since our sun's spin on its axis is counterclockwise, all spheres in our sun's Event Horizon revolve counterclockwise in orbit around our sun.

The black hole at the core of our sun spins counterclockwise on its axis, so every planet in orbit in the sun's Event Horizon revolves in a counterclockwise direction around our sun.

One black hole's gravity can trump another's.

Moons tag along with the planet they belong to as that planet revolves counterclockwise around our sun. But, the main orbit for a moon is the one it takes around its planet. The reason this is so is that proximity matters. Which gravity is closest to a moon? For a moon, its planet's gravity is the nearest gravity affecting it, so the moon revolves around the planet. Thus, the planet's gravity trumps the further away, yet stronger gravity of the sun.

Two different gravities: A black hole's gravity pull can guzzle light or hold matter in place.

The gravity of a black hole works in two different ways, depending totally upon a black hole's fullness.

Gravity for a naked black hole at the CREATION of matter phase works different than the gravity of a black hole inside the core of a planet or star during a LOCKUP holding phase.

Both these gravities have tremendous force, yet they act differently. The reason is that the black hole changes its duty as the black hole fills with matter.

Guzzling Gravity – starting from empty.

With a naked empty black hole, its gravitational vortex is dedicated strictly to sucking in light. Guzzling is the name of this process. A black hole is constructed such that its body spins around a central spiral helix which lengthens and shortens with each superfast powerful spin to perpetuate and accentuate a strong magnetic field. The black hole will use the power of guzzling gravity to attract and spin straight line light into matter during this CREATION of matter phase.

The gravitational force of the fast spinning black hole entices nearby rays of light. Rays are curved and slowed, then bent and spun acutely to the extent that the head of the light ray turns sideways and spins like a barber's pole thus creating a spinning helix which pulls and spins the rest of the tail of the light ray in behind to form an atom.

During the CREATION phase, light attracted by a black hole is slowed, bent and spun into matter.

Gradually, the Guzzling Gravity of the black hole attracts evermore light which is devoured as the hungry black hole jails more light into atoms of matter. The earliest formed atoms align with the black hole's helix to form a more powerful axis. As matter builds up, both the helix axis and the core body of the black hole increase in size.

This strong Guzzling Gravity during the CREATION phase has bazillions of tentacles leading out from compartments of the core bowels of the black hole. At the beginning of the CREATION phase, every single compartment is dedicated to guzzling, capturing and spinning light into atoms. But, once a tentacle compartment captures a ray of light, it has done its quota, meaning it can no longer perform guzzling. It must now dedicate itself to holding and LOCKUP. And, as more compartments fill they switch their dedication away from Guzzling Gravity to Holding Gravity.

Holding Gravity of a full black hole.

There comes a tipping point for the full black hole as its compartments become 100% dedicated to holding during LOCKUP. Thus, no further guzzling can occur. This lockup is so powerful that the black hole's Holding Gravity on its atoms extends out past the black hole's core to its surface to hold anything and everything from departing the surface. and flying out into space. For any living organisms on the surface there is difficulty moving away from the surface as all efforts to jump free or fly away result in a gravitational drop back down to the surface. It takes a tremendously strong effort, such as a space shuttle rocket, to escape the powerful hold of LOCKUP during Holding Gravity exerted by a black hole at the center of a sphere.

Now, suppose that the black hole at Earth's core is still attempting in vain to use its gravity to attract and spin more light into matter. But it can't as there is too much matter in the way. Us, being held on the surface and not flying off is perhaps just an accidental benefit of this attempt.

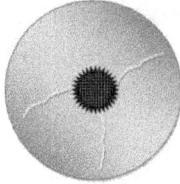

As the Black Hole grows larger, all tentacle compartments have captured sufficient rays of light to form a sphere (a moon or planet). Now, each of the compartments rededicate themselves to a new job, namely to Holding Gravity in order to hold spun atoms in place and prevent them from escaping.

For any planet, power of this gravitational hold extends out past its own surface into space to nearby moons in orbit. For a sun, gravity extends far out to the planets.

The rest of the life of a sphere at this full stage will be dedicated to Holding Gravity, maintaining a strong LOCKUP of all of its atoms for as long as possible. That effort usually pays dividends for billions of years during a LOCKUP phase.

LOCKUP in the body of the black hole, now turned planet, can last for billions of years. However, after billions of years of LOCKUP, as the sphere further ages, the compartments of the sphere tire and begin to lose control of their contents.

The first freed atoms can form into extreme heat at the core of the planet. Further along in time, a few billion years later, the planet can even swell to the size of a small star.

Even further along after another billion years, the planet

can lose control making its surface hot and liquid. Near the end of the LOCKUP phase, it can heat up and emit light as a sun. In the end, the sun can entirely fire up and finally explode as a Supernova. All light trapped escapes back to freedom as straight line light (ESCAPE), incredibly somehow escaping from the powerful gravity of a tired sun.

How light is able to escape our sun's powerful gravity.

At the start of our solar system, only naked black holes remained after a Supernova had destroyed the old system. The largest naked black hole of that new system attracted and spun the most light into matter. It was also the first naked black hole to gobble sufficient light to grow into an xx-large planet, and switch from guzzling to holding gravity. Possessing the most mass, its stronger gravity made it the central sphere of a new solar system which still lacked a sun.

But, it was also the first sphere to experience a loosening of atoms because it had been a total glutton and overeaten. Other black holes were still guzzling. These holes further sped a loosening of its atoms which became rampant with its surface firing up entirely turning it into a sun.

Thereafter, light rays freely left the sun's surface. Other black holes orbiting that sun soon ate so much of its light that they grew into moons and planets, entering phases of holding gravity. So, while our sun possessed sufficient mass to hold spheres in orbit, it lacked the necessary Holding Gravity to keep its atoms from unspinning back to original straight line light and escaping from its sunny surface.

Guzzling Time compared to Holding Time.

The time a black hole spends spinning light into matter is short compared to the time spent holding atoms in place.

Guzzling Gravity time duration (CREATION phase) may only be a few years, dependent upon the plentitude of the supply of light available to be spun into matter and also dependent upon the size of competitive black holes nearby.

Holding Gravity time duration (LOCKUP phase) may take anywhere from eons to up to fifteen billion years for a black hole which goes thought the entire cycle from naked black hole, to solid sphere, to large solid sphere, to liquid surface sphere, to small sun, to a larger sun, to Supernova death.

Several factors determine this eventual ESCAPE of light.

♦ A black hole always overeats during CREATION phase. It is always too greedy in its desire to devour light.

♦ Speed of spin of a black hole gradually slows over billions of years as it goes from a naked black hole to full sphere.

♦ Light trapped and spun as a ball inside an atom of matter always wants to escape back to free straight line light.

♦ Light will always be successful in its escape from its jail in matter, even if it takes eons or billions of years.

♦ Light deep within the central core of the black hole of a sphere will always loosen the soonest. This is the area of the oldest spun matter.

♦ Deep in the core, the first escape of light and heat begins, and that's where a fire starts up.

♦ This internal fire deep in the core always tussles its way outwards to the surface as the heat expands.

♦ This loosening and resultant fire always moves outwards from the core through fissures. This fire finally reaches the surface as molten lava. Billions of years later on, this fire can liquefy the planet's surface, making it into a fiery sun.

Speed of spin changes with the type of gravity.

During Guzzling Gravity (during the CREATION phase) the naked black hole demonstrates incredible spinning speed used to devour light and spin it onto matter.

This incredible spin speed can even affect other spheres nearby as there can be an extended draft out into space. Spheres in the direct vicinity of that draft are susceptible to a premature loosening of the atoms comprising their bodies. Nearby suns can experience a premature loosening from an attack by a black hole causing a sun into an early Supernova.

During Holding Gravity (LOCKUP phase) the spin speed of a black hole slows dramatically. As the mass of the sphere increases, so does the extension of its Holding Gravity away from the sphere to hold moons or planets in orbit.

During the last of the holding gravity (ESCAPE phase), when light is escaping from the surface of a sun, the rate of spin for the black hole at the sun's core slows even more.

Supernova affects the gravity of all involved spheres.

At the moment of any Supernova, the direction of spin reverses for the black hole at the core of the sun. This is caused by the whipping backlash from the instant release and escape of tons of atoms (for every action there is an equal and opposite reaction). In the end, this backlash of speed splits the axis core of the black hole into two parts.

The planets and moons of the solar system are released from the gravity hold of the Supernova sun. Spheres closest are obliterated (they have their own quick expansion and supernova as their atoms all burst into light). Some distant spheres of the solar system may survive and be driven from the blast area further out into space as solitary rogues.

It is during this replication phase where newly twined naked black holes receive the direction of rotation on their axis and the direction of the gravity Event Horizon which will be with them for the entire cycle of that particular black hole and for the sphere which will grow around that black hole.

Conclusion to the two types of gravity.

In this chapter we've seen that there is the gravity of a naked black hole which attracts and pulls light inward to be spun into matter. And, there is the later gravity of a full cloaked black hole where the black hole is hidden away from sight at the core of the sphere which it spun around itself.

In mature solar systems, black holes are cloaked away hiding away inside of full spheres. Their gravity during these long holding phases keeps atoms on the surface from flying off into space, and for holding other spheres in orbit.

Not till Trillion Theory do we see what gravity really is and what causes it - namely black holes. Till now, astronomers have seen black holes gobbling up light, but they had never discovered that the actual true purpose of black holes was as engines to build spheres and engines to control spheres.

Now in Trillion Theory, new terms *Uncloaked Naked Black Holes* and *Cloaked Black Holes* become important terms when fully understanding the gravity prevalent throughout the solar systems and galaxies of our cosmos.

As we discover more and more about black holes and how they build spheres, solar systems, and galaxies, we will discover more about gravity. So far scientists have never been able to manipulate gravity, perhaps someday they will.

Interviewer: "Something puzzles me. If the black hole at the center of a sun has greater gravitational holding power than a smaller sphere, how come light can escape from a sun?"

Author: "That's a tricky question. For the answer, we go all the way back to the start of that solar system, let's name that sun Henry. At the solar system's start, only naked black holes remained after a Supernova had destroyed the old system. As the largest naked black hole of that new system, Henry attracted and spun the most light into matter. Henry was the first black hole to gobble sufficient light to grow into a large planet, and switch from guzzling to holding. Possessing the most mass, Henry's stronger gravity kept smaller spheres in his orbit. So, Henry became the center sphere of that new solar system, a solar system which still lacked a sun. But, Henry was also the first sphere to experience a loosening of its atoms because it had totally overeaten. Other spheres held in Henry's orbit were still guzzling. These black holes further sped the loosening of Henry's atoms. This loosening became rampant. Henry's surface fired up into a sun.

Thereafter, light rays freely left Henry's surface. The black holes orbiting him soon ate so much of his light that they grew into moons and planets, entering phases of holding gravity. So, while sun Henry possessed the mass to hold the spheres in orbit, he lacked the necessary Holding Gravity to keep his atoms from un-spinning and escaping his surface."

Interviewer: "Thanks, I get it. Your explanations helped me see what causes gravity and keeps us glued to our planet."

Author: "I'm glad to see this is making sense. If TT is correct about a black hole at the core of every sphere, then this really changes how we view gravity."

CHAPTER 11
THE END FOR A
SOLAR SYSTEM

Supernova explosion of a sun and **Obliteration** of its solar system are the final acts in the roughly 15 billion years of one entire cycle of a solar system from its birth from battling black holes to the death of its spheres..

Trillion Theory then places Replication after the tail end of Supernova and Obliteration. Replication is where all the surviving naked black holes replicate themselves from one black hole into two. Thus, from the old destroyed solar system, every black hole which was at the core of a sphere now doubles into two new naked black holes. At this point, these new naked black holes begin the Rebuild phase of erecting a brand new solar system. This Rebuild phase is followed by the long Holding phase for the spheres. Our present solar system is a great example of a Holding phase.

Here is a rough timeline of one 15 billion year cycle from the start to the end of a solar system:

Rebuild > > > > (Formation of new solar system).

Holding (Control over spun matter for billions of years; the longest lasting phase and most visible to the eye)

> >

> >

> >

Supernova > (Death of the sun of the solar system).

Obliteration > > (Rest of solar system spheres destroyed).

Replication > (One into two. Easily the shortest phase).

All phases: Rebuild, Holding, Supernova, Obliteration, and Replication, are equally important to the recycle process.

Before examining the start of a new solar system, we look inside the old solar system to examine how it was destroyed.

Supernova - End of the cycle for an old star.

It's an incredibly brilliant sight to witness a star explode. While many of them happen across the vast expanse of our cosmos each day, you have to be in the right spot at the right time to catch a Supernova explosion in action.

Supernova of a star happens for two reasons.

Scenario one occurs most times. The sun has reached old age at roughly 13 billion years into its cycle. It experiences extreme difficulty in controlling its contents as an inordinate number of its atoms are unraveling and escaping back to freedom, as straight line light. As this process escalates, more atoms unravel at a hectic pace and frequently larger solar burst from the surface. The sun expands, providing an over-abundance of heat and light to its solar system. As the sun expands enormously, its heat engulfs the closest sphere held in its orbit. Then finally, the sun loses total control of its contents and explodes into a Supernova.

Scenario two is quicker: A naked black hole comes in close proximity to the sun; changes occur at a much accelerated pace. The naked black hole ferociously eats of the sun's light causing the sun's contents to unravel even faster.

At the Supernova moment, things turn totally violent as the sun explodes all of its contents out into its solar system. The sun's atoms unspin and fire outward as freed light. But, while the sun dies, the black hole at its core survives.

Obliteration – end of the cycle for an old solar system.

The brilliance of the Supernova of a sun is enhanced by the explosive Obliteration of most of its solar system. The explosion's outburst wave from the sun engulfs most of that solar system. Every nearby planet and moon in that system swells and melts, flash-evolving. The planets and moons lose control, unspinning septillions of atoms back to light. For any sphere to survive the long reach of Supernova it would have to be several billions miles away in a distant orbit. Unfortunately, those closer are not so lucky.

Black holes survive from Supernova and Obliteration.

The only part of the sun which survives the Supernova, or part of a planet or moon which survives Obliteration, is the black hole at their cores. A black hole cannot be destroyed.

In that graveyard of the old solar system, after Supernova is followed by Obliteration, there no longer is a sun or any planets or moons, except for perhaps a distant planet lucky enough to be in an outer orbit far enough from the blast to survive. Even so, the rogue may be flung far out into space.

There is little available evidence that the old solar system ever existed. All light which had been trapped for billions of years as matter in the spheres is now flashing outwards attempting a bid to escape the area. Or so that light thinks.

At this point, it's difficult for even a powerful telescope to hone in and clearly see what has occurred. For unknown, until TT, is that black holes resided at the core of every sphere in the old solar system and every one of those black holes survived the Supernova and the Obliteration. In the aftermath graveyard, all the surviving black holes are now uncloaked, naked, hungry, and intent on devouring light.

Interviewer: "Where does an old star go when it explodes and dies? Does it spread its material like bits of sand?"

Author: "Not like sand, rather like light. The material of the star departs in an explosive hurry. The Supernova unspins all its remaining atoms at one precise moment. The huge flash of light means a tremendous amount of light is released by the star back to straight line light which is then freed from captivity to travel the cosmos. However, not all of this light escapes. Most of it is re-spun back into matter by the naked black holes present in the area right after the Supernova.

However, while the star died, its black hole core survives. The backlash from the force of the Supernova will split that surviving black hole into two new black holes. Hence, there are now two x-large black holes to battle for control of a new solar system or perhaps two totally different systems."

Interviewer: "Yikes! How could a black hole core survive the supernova explosion of its star?"

Author: "There are two cosmic materials which can never be destroyed; they are light and black holes. Light can never be destroyed, only recycled. Light can be spun into matter; then billions of years later that matter unravels back to light. In $E=mc^2$, energy and matter can't be destroyed. But, they can be recycled back and forth across the $E=mc^2$ formula.

For a black hole, its composition must be very elastic and pliable to withstand the power of a Supernova. As well, that black hole inside of the occurring Supernova is losing tons of atoms from its core in mere instants. So, to protect itself from being destroyed, it appears that the black hole has gone so far as perfecting a splitting action (Replication) as a means of taking the brunt of the force of the Supernova."

CHAPTER 12
REPLICATION
AND REBUILD

Replication is a term used for the reproduction of one black hole into two new black holes. **Rebuild** is the rebuilding of a solar system by naked black holes after their old system was destroyed by a Supernova.
Trillion Theory (TT): Replication by black holes caused the growth of our cosmos over the past trillion years.

This is amazing that a naked black hole can replicate itself by splitting into two after surviving the Supernova of a sun and Obliteration of a solar system. Black holes are survivors, able to survive anything, and able to replicate themselves after the traumatic events of a Supernova and Obliteration.

TT uses the term Replication (making two new black holes by splitting the old black hole) in describing how black holes increase their numbers. This Replication feature of black holes is what has allowed our cosmos to start small at origin a trillion years ago and then grow exponentially with every new cycle to a present size of 73 quintillion stars within billions of solar systems within multitudes of galaxies.

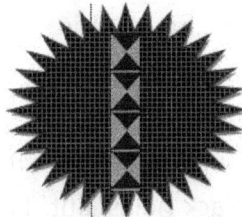

Spiral helix is an interior component running top to bottom inside of a black hole. It spins the black hole on its axis.

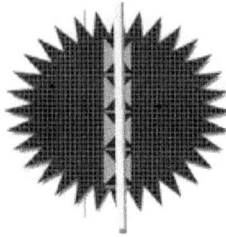

The helix splices and splits when a Supernova exacts its force on the black hole which was residing at the sphere's interior. This action splits the black hole.

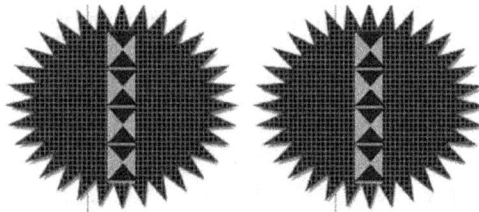

The one black hole splits and duplicates during Replication to form two new naked black holes. The force of this split drives the two new black holes away from each other.

Replication (the doubling process).

When a sun goes Supernova, the awesome explosion of matter all occurs in the blink of an eye. The sun jumps into an instant fast rotation, as the backlash from tons of matter leaving its surface spins it ferociously like a top. Now, as we know, for every action this is an equal and opposite reaction. At that instant, the surviving black hole at the sun's core will backlash violently. This backlash is so powerful that the helix and the body of the emptied black hole split and thereby replicate from one black hole into two black holes.

This replication could be perfect with the result being two near identical twin black holes. But, TT contends that such a powerful split would more often be imperfect, meaning one of the new naked black holes could be larger than the other.

Since there is the possibility of a size differential between the two new naked black holes, the smaller black hole could be taken into orbit by the larger black hole.

Also, this forceful split of a black hole imparts spin to both of the new naked black holes. Most often they will spin in an opposite directions to each other. However, according to TT Cosmic Law they both could spin in the same direction. (We see that galaxies in our cosmos are approximately 50-50 when it comes to direction of spin).

Of importance is that TT is able to explain why all cosmic spheres pivot (rotate) on their axis; the reason being that a spinning full black hole is at the core of every sphere. The direction of spin of the black hole is taken on by the sphere which the black hole built around itself.

The same phenomenal splitting and doubling transition occurs at the axis of every black hole at the core of every obliterated planet and moon in a solar system. In this fantastic rebirthing event, with replication by all the naked black holes of a destroyed solar system, there is now double the number of black holes compared to the old solar system. These newborn naked black holes are all poised to double the sphere population of the new system of that region. The next 15 billion year cycle of this solar system is set in motion. **Trillion Theory shows a definite progression of events occurring after a Supernova and Obliteration.**

• Double the number of replicated naked black holes results.

• Each new naked black hole is immediately competitive with each and every one of the other naked black holes in the area. It is survival of the fittest as the early birds (the two black holes resulting from the old sun) are larger and poised to win the battle for light in their particular area. That area is without a new sun as yet, and the moons/planets are just beginning to form as the black holes spin matter around their bodies. Gravitational forces see larger black holes holding smaller holes in orbit, setting an early configuration.

• Some light escapes the vicinity right after Supernova and Obliteration. This escaping light exits out into space, free to travel after billions of years being jailed up inside of matter.

• The rich get richer. The largest naked black holes have the upper-hand on all the lesser sized black holes. Brilliance occurs when some of the light can't escape and gushes back inwards towards the largest black hole. This is prolifically seen in one of the black holes where the old star existed, as that large early bird naked black hole is ready to recapture much of the light exploded away from the moons and the planets after Obliteration occurs. A person could feel sorry for light which spent billions of years in entrapment, praying for the day when its atom would unravel and allow escape, only for light to be quickly recaptured and jailed again.

Rebuild: Black holes build a new solar system.

Every solar system rebirth is as different and unique as a snow flake. When a smaller solar system gets obliterated, its surviving black holes will likely stay close enough together to rebuild into one double-the-size solar system. Whereas, when a sun is massive its Supernova is bound to drive and fling its solar system far apart; therein two far apart smaller solar systems would be the prospect.

Rebuild: Fight for light in building new spheres.

Right after Supernova/Obliteration, the region of space around all the new naked black holes becomes a war zone.

All the holes spin at their mightiest speed to cannibalistically snap light stands to spin into their core compartments.

In the new system, temporarily comprised of just naked black holes battling each other to consume the most light, the old graveyard in space neither remembers nor shows anything of the old solar system of sun, planets and moons. All that remains from the old solar system is double the number of naked black holes and light flung outward from the Supernova and Obliteration. There is little true evidence that the old solar system ever existed - recycling is efficient.

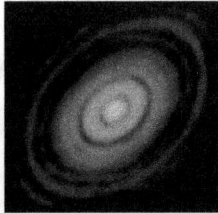

The Rebuild of a new solar system begins. All the naked black holes forming up the new system revolve in orbits around the largest dominant black hole at the system's hub. Although black holes are an ebony color, white appears throughout the area as light is drawn in by the black holes which capture and spin light into new spheres for the new solar system.

Rebuild: Surviving black hole of a sun takes dominance.

Dominance normally comes in the form of the largest (XL) of the split black holes from the old sun taking charge of the organization of the new solar system.

This XL will initially be thinner than its parent. But, black holes are very elastic entities with capabilities of expanding their helix and compartments. So, XL will quickly get back to being even larger as its gobbles light to increase the length and girth of its spiral axis. This process also expands and increases the size of its compartments and their numbers.

The XL black hole has a distinct advantage having done Replication before any lesser black holes at the cores of the moons or planets. The XL scoops up light from those smaller neighbors when they explode during Obliteration.

XL has the upper hand to become even larger compared to the lesser black holes as it spins the most light to add to the mass necessary to hold all lesser black holes in orbits. This battle for light will decide the new head honcho of the new solar system. XL gobbles the most light, gluttonously filling up, becoming the first of the black holes to complete a body. XL will be the first sphere to change from Guzzling Gravity (light capture) to Holding Gravity (atoms controlled). Thus, this XL is the top candidate to have its atoms loosened and one day evolve into the solar system's new sun.

Of course, each new solar system is totally unique, no two snowflakes exactly alike. With each new solar system, each planet/moon has its own size, mass, composition, climate, surface activity, orbit, and speed of rotation on its axis.

However, how a solar system (or a galaxy) forms a certain way is more than just pure happenchance. Definite specific Cosmic Laws exist between black holes, suns, planets, and moons, providing similarities in organizational makeup.

Black holes, those building engines of the spheres of our cosmos, are guided by a specific set of Cosmic Laws when they form solar systems and then galaxies.

However, don't ever get the idea that naked black holes are always nice guys. They are absolute ravenous battlers when fighting against other black holes over the attraction and consumption of light - all done under the guidance of Cosmic Laws pertaining to black holes and spheres.

Interviewer: "Is the new solar system or systems formed after the Obliteration different than the preceding one?"

Author: "The obvious answer is yes. Each new solar system is unique, no snowflakes exactly alike. Therein, the number of moons around each planet can vary from one solar system to the next. Also, every planet/moon has its own size, mass, composition, climate, surface activity, and specific orbit.

However, how a solar system (or a galaxy) ends a certain way is more than just pure happenchance. Specific 'cosmic laws' exist between black holes, suns, planets, and moons. These laws provide definitive similarities in organization.

Allow me to define the cosmic laws as having come from both 'the rules of co-operation' and 'the rules of battle.' Black hole at the core of full spheres usually co-operate in solar system and galaxy settings. Whereas, naked black holes have battling rules of engagement when fighting each other for light and for best positioning within a solar system or galaxy. These battle situations are akin to pigs at a trough; contests where the strongest win. Cosmic laws prevail."

Interviewer: "If black hole can replicate, are they alive?"

Author: "Yes, but *alive* in a different way." To be considered *alive* means to meet *alive* criteria: have mass; use nutrients; grow; reproduce. Well, black holes meet all criteria. In TT, it is also shown that supermassive black holes can change their structure thereby strategizing how they will control a galaxy for hundreds of billions of years. Are they sentient?"

Interviewer: "Most interesting. But, I'm not a TT believer just yet. Although, if someday proof is found, then TT may place humans on the threshold of solving our universe."

CHAPTER 13
DETAILING COSMIC LAWS

Inside of TT there are Cosmic Laws. It is important that these Cosmic Laws be categorized as these laws are the links which tell us why many similarities exist between certain types of spheres, and why there are similarities of organization between solar systems, and why there are also similarities of organization between spiral galaxies.

These similarities result from the consistent Cosmic Laws (rules of engagement in cosmic actions) which regulate the activities and relationships of black holes and also spheres. Classifying these various Cosmic Laws begins with black holes and then proceeds to spheres, as follows:

▪ naked black holes are of small, large, x-large, massive and supermassive size.

▪ spheres are of small, large, X-large, and massive size.

General Cosmic Laws for naked black holes:

• A naked black hole has an insurmountable desire (mission) to devour light, and a directive to spin that light into atoms of matter in for the purpose of building a sphere.

• A naked black hole can devour light in amounts relative to its naked empty size. And, a naked black hole always wants to increase its size from one cycle to the next.

• An empty black hole always eats light faster and in greater amounts than a full black hole such as a planet.

• A naked black hole, because of its spinning power, creates gravity extending strongest outwards from its equator.

Cosmic Laws for SMALL naked black holes:
▪ A small naked black hole devours less light to spin into matter around itself than a larger naked black hole.
▪ A small black hole exerts less gravitational pull.
▪ A small naked black hole has difficulty winning battles for light or using its gravitational pull against larger opponents.
▪ A small naked black hole spins only tight atoms around itself, thereby building a hard surface.
▪ A small naked black hole spins a small sphere of matter around itself. Therein, a small black hole, once full, would be at the center of a moon or small planet in a solar system.
▪ A small naked black hole would spin too small of a sphere around itself to ever have a chance of evolving into a sun.

Cosmic Laws for LARGE naked black holes:
▪ A large naked black hole can devour more light to spin into matter around itself, than a smaller naked black hole.
▪ A large black hole will overpower smaller black holes in the battle to spin light into matter.
▪ A large naked black hole exerts a stronger gravity pull.
▪ A large naked black hole's gravitational power reach can hold smaller spheres in orbit.
▪ A large black hole spins looser atoms, making it the fitting center for a larger sphere such as a large planet.
▪ A black hole of larger size has better odds of someday evolving from a moon to a planet to a star.

Cosmic Laws for EXTRA LARGE (XL) naked black holes:
▪ An XL naked black hole can devour more light to spin into matter than a small, medium, or large naked black hole.
▪ An XL naked black hole will overpower smaller black holes when it comes to devouring light to spin into matter.
▪ An XL naked black hole exerts a much stronger gravity pull.

▪ An XL black hole, because of its greater gravitational reach, can hold small and large spheres in orbit around itself.

▪ An XL black hole spins the loosest atoms as a body, which makes its surface more gaseous and less solid.

▪ An XL naked black hole has top odds of further loosening and evolving into the sun of a solar system.

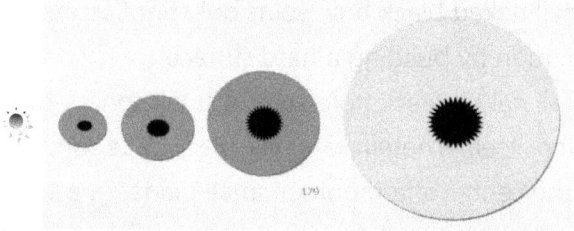

Black holes have varying sizes regardless if they are naked or if they already have a body of matter spun around them. Left to right above: naked; small at center of small moon; medium at center of a medium sized solid planet; large at the center of large planet; XL at center of a star such as our sun.

Cosmic Laws for MASSIVE black holes:

▪ A massive black hole possesses a superior size and gravity advantage of all the lesser black holes in its area.

▪ A massive black hole uses this advantage in the building of an x-large solar system around itself.

▪ A massive black hole builds a massive sun around itself.

▪ A massive black hole inside of a massive sun has superior gravitational mass to also hold other solar systems in orbit forming a small galaxy around itself.

▪ A massive black hole inside a massive sun will tire of going Supernova and have to rebuild its small galaxy.

▪ A massive black hole has a goal to become supermassive and one day become the center of a gigantic galaxy.

Cosmic Laws for SUPERMASSIVE black holes:
(supermassive black hole at the center of a galaxy).

▪ A massive black hole's top priority is to become the center of a galaxy, therein becoming a supermassive black hole.

▪ a supermassive black hole's goal is to control its galaxy for a period of up to hundreds of billions of years, thereby overcoming the normal 15 billion years allotted for a normal solar system or a tiny galaxy.

▪ a supermassive black hole deploys new strategies with its structure in order to break the rule of 15 billion years.

▪ a supermassive black hole deploys new strategies allowing it to live much older, up to 500 billion years or more.

▪ a supermassive's structural changes allow it to remain as a supermassive black hole and not have to grow into a sun and someday go Supernova destroying its galaxy.

▪ a supermassive's black hole doesn't need to take on the sluggish rotation which would occur if it became a sun.

▪ a supermassive black hole's structural changes allow it to pull in tons of light, therein able to hold large suns around itself (seen as a halo), and pass this light right through its hollow helix pluming and jettisoning out its poles.

▪ a supermassive's black hole utilizes the gravitational mass of the suns which it holds in a cluster around itself to assist with the holding and spinning of the entire gigantic galaxy.

The supermassive black hole at the hub of a spiral galaxy is surrounded by a halo of stars. Tons of light pulled is inward by the supermassive and jettison out its top and bottom.

Now, here are the Cosmic Laws pertaining to spheres; these laws deal with the situations occurring after black holes have built spheres around themselves.

General Cosmic Laws for spheres:

▪ A sphere with a full black hole hidden and cloaked in its core has all of its contents held in place by the spinning gravitational force of the black hole at its core.

▪ A sphere exerts gravitational pull around itself relative to its mass and the size of the black hole at its core.

▪ A sphere with a larger black hole at its core and more body mass extends its gravitational pull further outwards.

▪ A sphere with a black hole at its core extends the pull of its gravity strongest out along its equator.

▪ A sphere formed around a larger black hole can hold more spheres in its outer orbits; while smaller spheres are held in orbit by larger spheres.

Cosmic Laws pertaining to sphere population:

▪ The black hole population and the star population of the cosmos doubles with each 15 billion year cycle.

▪ The black hole at the core of the central sun of each solar system gains in proportion to the smaller holes at the cores of planets and moons.

▪ Space grows ever larger with the billion to one ratio of space left behind after light is spun into matter.

▪ Expanse of the entire cosmos accelerates outward with each new solar system being pulled towards the direction of the static ocean of light on the outer perimeter.

▪ The old graveyard provides no fossil evidence to indicate that the old wiped out cycles of solar systems ever existed.

Cosmic Laws relating to SMALL SPHERE size (example: a small moon or planet in our solar system):
- Small spheres are controlled in orbit by larger spheres.
- Small spheres end up being moons of larger spheres.
- Small spheres are most easily pulled close to their sun.
- Small spheres have denser packed bodies, concentrated with tight element atoms. Their surfaces are harder.

Cosmic Laws relating to LARGE SPHERE size (example: a large planet in our solar system):
- Larger spheres have more gravity for the holding of other spheres in orbit.
- Larger spheres are able to stay further away from a sun.
- Larger spheres are less likely to be pulled close by a sun.
- Larger spheres have looser gaseous surfaces.
- Larger spheres in size-mass exert stronger gravity pull.
- Larger spheres have a greater possibility of evolving and growing into a star.
- Larger spheres (in far out solar system orbits) might survive a Supernova by their sun.

Cosmic Laws for EXTRA LARGE SPHERE size:
(example: sun of our solar system; and stars).
- XL spheres have an extra-large black hole at their core.
- XL spheres easily evolve over small opponents.
- XL spheres have more gravity to hold spheres in orbit.
- XL spheres can become the central sun of a solar system.
- XL spheres, when they were being formed, had the loosest atoms of matter spun into their bodies.
- XL spheres generally have a liquid-gas body.
- XL spheres are first to lose control of contents.
- XL spheres (suns) can mushroom and explode.

Interviewer: "Our universe seems to be quite a lot the same in various quadrants. Is this true?"

Author: "Yes. We will see definite similarities between solar systems across our entire cosmos. This is the result of the consistent cosmic laws (rules of engagement) regulating the activities of black holes and spheres. TT calls them rules of engagement because naked black holes (the light eaters) and the spheres which they build around themselves, always have an ongoing battle and relationship with the other black holes and spheres in their solar system."

Interviewer: "I admit that cosmic laws help me gain a more thorough understanding and a greater appreciation for TT."

Author: "The same has happened to me. I now appreciate our entire universe more than ever.

For the next moment, I'd like to open up discussion on a touchy subject, beginning with a question: Is our purpose as humans to simply enjoy, or to work hard, or to discover? Plus, we could add many more possible reasons behind the true purpose of human existence. But, I suggest this: Who or whatever did the cleverness of the inventions of light and black holes, or wrote the Cosmic Rules which the spheres, solar systems, and galaxies of our cosmos follow, also laid out breadcrumbs for us to discover. Some of these clues are: fire (partial return of matter to light); lightning (partial return of matter to light); a rainbow (bending of light suggesting light can be spun); a nuclear blast (release the tremendous power of an atom back to light); our sun (the slow decay of atoms as they unravel and return to straight line light); a Supernova (the instantaneous unraveling of a star sending all of its matter back to light). Clues to help us find our way."

CHAPTER 14
THE INVENTION
OF SPACE

Trillion Theory wagers that everyone else rationalizes space far differently than does TT. Everyone seems satisfied with how Big Bang depicts that all matter exploded across space at the supposed cosmic origin. Such acceptance would need to use the premise that space was in place awaiting the occurrence of the supposed Big Bang.

But, Trillion Theory says, "not so to Big Bang, and not so to space being in place ahead of the origin of our cosmos."

Everyone will likely reply, "Space had to be there first; we see space today and it is quite evident that our cosmos sits in space. Space is the 'box' which cosmos originated into."

However, TT advocates that space was not there first. This neo out-of-the-box TT concept throws people for a loop.
An easy experiment depicts space as TT sees it.

Spread out iron filings to completely cover a table. Think of the filings as TT's static ocean of light. Toss 20 magnets into the table full of iron filings, being sure that there is some distance between each magnet. Think of magnets as symbolizing black holes. Of course you will find that the iron filings are drawn towards and adhere to the 20 magnets. Think of iron filings stuck to the magnets as matter forming atoms around a moon or a planet centered by a black hole (a magnet). On the table, back where iron filings used to be would be nothing. Think of this emptiness as SPACE.

Each magnet, symbolizing a moon, a planet, or a sun now weighs more with the adhered heavy iron filings. Whereas the empty vacated barren table areas now weigh nothing, (like empty space). Thus, TT shows space as a void emptiness left behind caused by the formation of spheres.

When black holes spin light into matter, all of the light's properties end up as spun atoms of matter. Space is just an empty left-behind byproduct. That's how space originated in TT. Now having been built, space holds the contents of our cosmos and act as this super highway for light.

Definition of space in Trillion Theory.

TT defines space as the empty nothingness byproduct left behind when light from the static ocean of light was spun into matter as spheres. Space is the empty footprint, which spun light left behind when light was spun into a sphere.

TT puts it this way: first there was the static ocean of light; black holes were introduced into the ocean of light; these black holes took static light from ocean and spun that light into matter; space was the empty byproduct left behind when static light from the ocean vacated its original area.

We can also state that space is that nothingness that the spheres of our cosmos seem to be sitting in. Space, that vast emptiness, takes 99.99999% of the room in our cosmos. Galaxies, stars, planets, moons, comets, meteors, asteroids, and gaseous clouds take up less than 1% of the room.

Space offers no resistance as light from stars travels for billions of miles and years effortlessly through it. Thus, space can best be summed up as the nothingness housing certain matter (spheres) and a highway permitting free passage. Space is empty, black, weightless, odorless, and frigidly cold (all the nothings that light left behind when it was spun into matter). Neo Trillion Theory here depicts how space actually came to be as a left-behind byproduct of light's departure.

Back to the origin of the cosmos according to TT.

All that existed at the origin of our cosmos was the static ocean of light. When the first black hole was introduced into the static ocean of light, that black hole freed static light from the ocean of light and spun that light into matter.

Two things immediately happened: firstly, light spun into matter which created the first sphere around a black hole; secondly, space was created and left behind as a footprint where light used to be. Then, as more black holes populated our cosmos, space increased in size, expanding outward.

Occupying our cosmos was the endless static ocean of light.
Then, a black hole was introduced into the static ocean of light.
It devoured light spun into matter to form a sphere around itself.
All properties, such as weight and heat carried in light's tool box,
went with light into the matter of the body of the black hole.
The large dark area between the black hole and the static ocean
of light on the outside became empty space.

Free passage on an express space highway.

Light travels through the final frontier of space from one distant star to another part of our cosmos. It is given this total free passage as space never charges any toll. There seems to exist this nice covenant between the passing light and space, the express highway provider. Space is the able servant of the cosmos, while the masters of our cosmos are light and black holes.

Light can travel through space for billions of miles and many light years. If the cosmos is a trillion years old, some light may have been traveling for that length of time. More likely though, certain light may have gone through many cycles of spinning into matter and unspinning back to light.

In TT, if we traveled to the outer edge of space, a person would see the endless static ocean of light, like a wall of stacked lumber. For all that existed prior to the formation of spheres was this solid ocean of light. As cosmos expanded, more light was used, leaving more interior empty space.

Formula of matter and space. Why space is so vast?

A formula could be used to find a ratio between spheres and space. Why is space so vast between stars?

Space is much more famous for what it doesn't contain, than for what it does. For, space is simply a void emptiness existing without weight, material, or warmth.

However, room-wise space is prominent. While there exist billions of spheres and galaxies, empty space occupies by far the greater area: stars only occupy .0000000000000001%; while space takes up .9999999999999999% of the available room in our cosmos. Comparatively speaking, matter is so densely packed, that one atom of matter spun from light leaves one trillionth that amount of space behind.

1 atom of matter = 1,000,000,000,000 or so area of space. Using our solar system as an example, we sum the volumes of every sphere and come up with an overall solar system volume. Then, we take that volume and compare it to the volume of space left behind when our system formed. This is a most hair-raising calculation as it is 4.2 light years through space from our solar system to the nearest Centauri System.

Another more micro way to calculate the ratio of matter to space is by examining an atomic explosion. During such a nuclear explosion, the nucleus of a shattered atom releases tremendous pent up energy. This immense energy takes up a vast area in ratio to the spec of matter, a million, billion, or trillion to one. This nuclear release of energy (ESCAPE) is a mirror reverse of the spinning of matter (CREATION).

Now, Trillion Theory contends that black holes played the vital role in the determining the size of space. The distance between stars is so immense and space is so vast because of the disproportionate ratio which exists between the small sizes of the spheres compared to the billions of miles of space left behind as empty byproduct from light being spun into matter by black holes. This ratio is staggering.

The shape of the space of our entire cosmos.

Trillion Theory shows that the general shape of the space of our entire cosmos is quite linear like a cucumber. Early in our cosmos, black holes could have been eating along a very linear plane. But, black holes can tilt away to more strategic angles from the linear plane when battling for light against other black holes. Therein, the space of our entire cosmos might be fairly linear overall, but not perfectly flat.

Black holes are shown eating light from the static ocean of light inside of the space they created. These black holes could have been eating in any swiveling direction along a fairly linear plane. But, the black holes didn't have to stick to an exact axial tilt, thus, our cosmos in not perfectly linear. Nor is our cosmos anywhere close to spherical as a Big Bang would predict.

Space touches the static light ocean on the perimeter.

Question: 'How is space so empty, couldn't light from the outer ocean of light simply flow inwards to fill up the space?'

A clue came from a news article released in September of 2013. The article entitled 'Scientists Freeze Light' showed another of light's yet-undiscovered properties. You see, light has so many different speeds: speed in a straight line; a slower speed when bent; slower still when passing through thicker mediums; slower again passing through diamonds; and slower again when spun into matter inside of an atom.

These scientists discovered for the first-time that light can actually be taken to a zero speed by freezing it. Scientists froze what is regarded as the fastest known thing (light).

The feat was not anything easy like placing light into a freezer drawer. Rather, scientists converted light coherence into an atomic coherence as a crystal and shot laser light through and trapped a laser beam for a full minute at a standstill, thereby halting light's fantastic speed.

This discovery gave Trillion Theory more clout by showing why the static (frozen-like) ocean of light doesn't just flow light into empty created space. The reason is that the ocean of light exists in a type of static frozen state.

So now, we revisit the onset of our cosmos a trillion years ago when our cosmos began with the ocean of light. The ocean of light was filled with static light (not in motion), sort of just frozen in place. Yet, frozen here does not refer to the temperature 'frozen.' Rather it refers to a unique means of suspending light so that its speed is zero and it sits and waits at rest. The ocean of light was at a standstill.

This outer ocean of light remains frozen forever unless its light is pulled away by the powerful spin of a black hole.

Remember in TT how the first black hole(s) at the origin of our cosmos attacked the static ocean of light, loosening and devouring it. Obviously, the tremendous spinning speed of a black hole loosened the light being held at a standstill. Once that light was jerked from the ocean, it immediately sped up. But, this freedom to fly was cut short for this light as it was quickly spun into matter inside the black hole.

In reality, the black hole would have eaten deeper into the static ocean of light creating an empty dark void cavern of space within the ocean.

Today, space has expanded to incredible limits as trillions of black holes have approached the light ocean and eaten from it, each time digging further into the ocean, expanding the void, and leaving larger volumes of space behind. Light not pulled from the ocean of light has remained static at zero speed for the past trillion years. An inexhaustible supply is still waiting to be utilized to build an ever-larger cosmos.

So, space is the empty byproduct which light laid-out as a future highway for itself to use whenever it escapes from a sphere. Various parts of the space highway has been utilized for hundreds of billions of years as a travel means for light rays, spheres, galaxies, comets, humans, and perhaps even aliens. This gigantic space highway has expanded each time more light has been drawn from the static ocean of light.

Interviewer: "Tell more about the scientists who froze light!"

Author: "The 2013 article entitled 'Scientists Freeze Light' gave us another of light's yet-to-be discovered properties. Light has so many different speeds: speed in a straight line; a slower speed when bent; even slower when passing through thicker mediums; slower again passing through diamonds; and slower again when spun into matter inside of an atom.

Now, these scientists had discovered for the first time ever that light can actually be taken to a zero speed by freezing it. Scientists had frozen what is regarded as the fastest thing in our universe (that we know of at present). The feat was not anything easy like placing light into a freezer drawer. Rather, scientists converted light coherence into an atomic coherence as a crystal and shot laser light through and trapped a laser beam for a full minute at a standstill, halting light's fantastic speed.

So hurrah, this gave TT what I had been looking for as to why the ocean of light doesn't just flow light into empty created space left behind by light spun into matter. The reason is that the light ocean exists in a static frozen state."

Interviewer: "Also, it's amazing that space is getting larger?"

Author: "Back in this book's preview, TT provided an analogy between space and a mining operation. Miners burrow deep into outer walls of a mine, reaping ore, while at the same time ever-extending the overall space of the mine. In similar fashion, black holes burrow into the walls of the static ocean of light, breaking free strands of light and quickly spinning them into matter. All the properties of those vast amounts of light go into matter, while the size of empty space increases a million fold inside the depleted hollow area of the ocean."

CHAPTER 15
TIME'S COSMIC
INVASION

'Time is an illusion.' Albert Einstein.

Many people have taken a stab at defining time, such as 'Time is an overseer of all things.' TT presents a fresh new look at time and how it came to be a feature and so-called overseer of our cosmos.

Time is a tricky subject. It marches on unabated. On Earth, billions of people each day get to share the ticking away of time. Of course we often get warped or skewed views of time. When we wait impatiently, time drags endlessly. Yet, at the end of a long highly competitive day, it is difficult to recall back to the morning. It seems like days have passed.

TT definition of time is far different than ever proposed by anyone. It includes: what time is; how it was imported; and where it exists or doesn't exist inside of our cosmos.
TT definition of time.

Trillion Theory definition of time: 'Time is imported inside of our cosmos right along with the spinning of light into matter. The measurement of time is the duration that light stays confined, spinning as matter below the speed of light.'

Time is a concept invented specifically for our universe. At the origin, there was only the static light ocean and time was nonexistent, but locked up inside of the waiting light ocean. Light was specifically designed and hidden inside of the zillions of strands resident in the ocean of light, waiting to be imported, become active, and invade our cosmos.

Not until a black hole took light from the static ocean of light and spun it into matter did time begin in our universe. Importation occurred nearly one trillion years ago as the first matter was spun and our universe's clock began ticking. At the moment of that spin, time imported into our universe, carried here inside of light's tool box. Time was an imported derivative brought into our universe by light and let out of the genie's bottle when that light escaped the ocean of light and was spun into matter around a black hole sphere.

Think of a ray of light traveling through space where it is able to reach the full speed of light where time stands still. Light carries in its tool box a quality known as 'time' which stays put until when that ray of light slows below the speed of light. When slowed, a ray of light sees time leap from its tool box and invade that specific area of our cosmos.

When a ray of light gets spun and trapped into atoms of matter by a black hole, that atom will spin for eons below the full speed of light. Time will exist within and around that atom until that light escapes and returns back to the full speed of straight line light. At full speed, time goes back into light's toolbox and time seeming once again stands still.

How time is measured in Trillion Theory.

At the full speed of light, time does not tick, basically standing still. It was Einstein who first showed this feature of the speed of light. If a spacecraft left Earth at the full speed of light, the people on the craft wouldn't age at all living in a no-time zone; whereas people on Earth would age in a zone which operates well below the speed of light.

Therein, time is calculated as the duration during which light moves below the full speed of light.

The starting point for time in Trillion Theory.

At the point just before the beginning of the building of our cosmos, even before time had been let out of the genie's bottle, all which existed within our physical cosmos was the static ocean of light extending endlessly in every possible direction. At that point, time did not exist in the ocean of light. However, inside that ocean, every single static light ray held in its tool box the stored the time feature.

With the introduction of the one initial naked black hole into the ocean of light, it did the duty it was programmed for. Deploying incredible spinning speed, the naked black hole broke chunks of light free from the static ocean of light, attracted it, pulled it in, and then spun the light into matter.

This breaking light away from the ocean of light was the beginning point for time in our cosmos.

A singular naked black hole broke the first rays of light from the ocean of light at the origin of our universe.
Time entered our cosmos accompanying the light ray which was quickly spun into atoms of matter.

As that first ray of light broke free from the ocean of light, it attempted to go from zero to light speed. But it never succeeded. That ray was quickly attracted, slowed, bent, and methodically spun into atoms of matter inside the belly of the naked black hole. At that moment, time first began. The duration during which that ray of light would spin trapped inside of an atom of matter defined and measured time.

For that first instance, time only existed inside of and in the direct gravity area surrounding the black hole. As more black holes spun light into matter, time spread further afield.

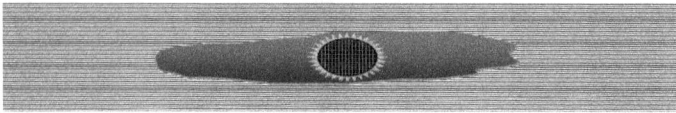

As the naked black hole ate from the static ocean of light, it spun light into matter around itself. That spun light moved slower and thus lived with a duration measure known as time.

When and where time exists in our cosmos.

Over a trillion years ago, TT states that time didn't yet exist within our cosmos. The ocean of light held light rays capable of carrying the time feature upon their release.

Once freed from the ocean of light by a spinning naked black hole, light brought time along with itself in its tool box to become active when that light would be spun into atoms of matter. (Einstein calculated: time stops at the full speed of light at 186,282 miles per second). But, TT states that time will begin to tick whenever light's speed slows.

Time doesn't exist in (1) the static ocean of light. And, time doesn't exist in (2) empty space or when a light rays travels at the full speed of light through empty space. However, time does exist in (3) where a black hole has broken light rays away from the ocean of light and spun those rays into matter around itself. Time is imported by spun light into the area of the sphere (3).

For whom was time created.

Here we need to think as inventors, and how the feature of time would have been designed to benefit more than one group inside of our physical universe. Time was ingeniously designed to function differently at various speeds of light.

For any planetary setting which might be habitable, the slowing of light inside of matter involved time which was designed to tick away at a normal pace so that the planet's occupants could experience the feature known as time. In this setting, growth and aging were fashionable.

But for explorers or caretakers for whom traveling our cosmos was a necessity, inventiveness had to overcome the challenge of the absolute vastness of space. Thus, time and aging could be nicely tucked away at the full speed of light.

Time on Earth. Tickety-tock.

On Earth, we don't fully appreciate the sophistication of this concept called time. For, time is much more than simply a clock ticking on the wall. Time is the measurement of how long light stays inside of matter in a spun state (below full light speed) as it counts down the duration of its captivity. The speed of light slows when bent by gravity, then further slows when passing through substances and further slows when confined and spinning inside an atom of matter.

Stated again in TT, time is defined as the measure of the duration it takes for light which is spinning as an atom of matter to loosen and unspin to escape back to free straight line light. Therefore, time exists whenever and wherever light slows below the full speed of light when encountering the gravitational pull from a black hole or a full sphere.

Time begins in a particular area of our cosmos when it is released from light's toolbox. This 'time release' occurs when light is attracted and slowed by a black hole. Since all fully spun spheres such as planets, moons, and suns have a black hole at their core, they all experience and live in a time zone.

Light attracted by a black hole is spun into an atom of matter. The various traits of light make up the various parts of the atom. The calculated duration which that strand of light stays trapped as a spinning atom is known as time.

Where time does exists and doesn't exist in TT.

Trillion Theory shows that time is specific to areas where light is slowed below its full speed. Time DOES exist where light has been spun into matter inside moons, planets, stars, and galaxies. So, time exists where atoms exist and where strong gravity slows, bends or spins light below top speed.

Whereas, time DOESN'T exist where light travels at full speed in blank empty space, but light can import time there.

Here is an example situation of how to view this: Suppose that we were inside a starship traveling at the speed of light through empty space. Even though we are matter, we are traveling at the full speed of light so time stands still for us. Suppose we passed another starship traveling at a tenth of light speed. Persons on the slower starship would experience time as they are moving below the speed of light. The same area of space for both, but the slower speed imports time.

We thereby deduce that space is not the carrier of time. The true carrier of time is a light, inside in its toolbox. Slow a light ray and time begins to appear in that area of space. Spin light into matter inside of a black hole and time is imported for eons of time into that particular area.

Therein, time doesn't exist in absolute unoccupied space. For time to exist in empty space, it must be imported into that area of space. Even the entry of a ray of straight line light traveling at the speed of light does not yet import time. It only imports a possibility of time. Take a light ray traveling through space at the full speed, time stands still aboard the light ray, even though the ray packs time in its tool chest. For time to start, that full speed light ray needs to encounter a gravitational station to make it drop below its full speed.

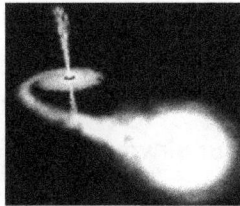

On a sun, time exists as light spins inside of helium atoms. When light escapes the sun and travels through space, time slows as the ray approaches the full speed of light, and time stops fully at the full speed of light when time hides away back inside light's tool box. However, when the ray approaches a black hole (or sphere) and slows, time is released from light's toolbox.

Time inside of a naked black hole.

We can say no, and then yes. Time didn't exist inside of the first naked black holes a trillion years ago. Any new naked black hole tossed into the static ocean of light was barren of time. Totally naked black holes are not carriers of time - light is. Thus, light imports time into a black hole.

Therein, time imports into the area surrounding a black hole once light is slowed, bent and spun into atoms by the black hole. Time is just a feature released from light's tool box when light slows and also when it spins into matter.

Invention of specialty item known as time.

It appears that time is a specialty item created solely for this universe. Time is carried by light and sprung into action from light's tool box and imported into an area time zone when that light's speed is slowed and also spun into matter. On the other end, when spun matter unravels and escapes back to straight line light, time converts back to hide inside light's tool box. Time can hide away while light travels at full speed though totally void space (a no-time zone).

Would time exist outside of our physical universe?

Tough query. It is difficult to say one way or the other with any certainty. Trillion Theory states that time is simply an extraordinary feature invented and created specifically for our universe and its spheres. Therein, time would likely not be ticking away outside of this universe. Why would the inventors want to age if they didn't have to? Therein, it's quite possible that our physical universe was created as a destination oasis for any desiring visitant to experience the peculiarity known as time. This peculiarity of time brings with it the notion or feeling that everything we experience has a beginning and end, time makes that so.

If outside this universe is in a no-time zone or realm, then the inference of a beginning or end might not be needed. Without time, there may be simply a continuance, with no beginning or end; or some manner of moving forwards and backwards through events.

An easy clue.

To assist humans to be able to grasp the concept of time, a clue was placed for us right inside of an old man-made timepiece. It is known as the winding coil. When a winding coil clock is totally unwound and lying dormant, it doesn't measure time. However, when the coil of the clock is wound tight, time is then measured while the coil slowly loosens.

In the clock, the key winding apparatus acts like an axis and spins the coil tight around itself. Time is measured by the duration taken for the coil of the clock to unwind. Then, a human hand must come along and wind the apparatus.

Now, compare an atom. With atoms of matter, time is measured by the duration that light slows and stays spun inside of matter. Atoms conveniently have a built-in rewind apparatus (mechanism); a helix pump tightens the atom's coil whenever it loosens. For eons the atom spins.

Time is a measure of the duration during which light's speed slows when spun into matter inside of an atom.

Unlocking more of light's incredible properties.

But, remember that light always eventually escapes from the atom and its time prison, even if it takes billions of years. Light will escape to return to travel though space at the full light speed where time stops aboard the ray of light.

In this chapter, Trillion Theory has once again shown new properties attributable to light. Each year light plays an ever increasing role in Earthly inventions. Will we further discover that light can be pushed past the full speed of light? If time slows and then stops at the full speed of light, would it move backwards at faster than the full speed of light?

Interviewer: "TT's depiction of what time is, and what causes time to occur, definitely gives me a fresh view of time."

Author: "No doubt time is a multi-dimensional concept. It's as tricky as tricky can be. It marches on one-directionally unabated around us. Seven billion people on Earth get to enjoy and share the ticking away of time each and every day. Perhaps our universe is a destination resort which we travel to in order to experience the exhilarating feeling called time.

TT's depiction of time is far different than anything ever written. Also, TT's proposal that time is imported into our cosmos is idiosyncratic. At the origin of our cosmos, time was initially stored inside the light stands in the vast ocean of light. When, some of this light was broken free by a black hole, those light rays carried time in their tool box. Then, when this light was spun into matter, time leapt from the tool box and began ticking in that particular spatial area,"

Interviewer: "I'm having difficulty understanding why space doesn't have time, yet space is part of our timely cosmos."

Author: "Time is a tough topic to get straight in one's head. The best place to start is by saying that space on its own can't possess time; however time can be imported into any zone of space. Try thinking of space on its own as a pure emptiness, totally void and bereft of everything, all heat, light, weight, odor, and time. Then, when a light ray passes through an area of space, it brings with it the possibility of time. Slow the light ray to below the full speed of light and time leaps into action. Spin light into matter in an area of space, and time will be the ticking clock measuring the duration during which light stays stuck as atoms of matter while spinning well below the full speed of light."

CHAPTER 16
NEW PARADIGM (ToE)

W **hat is a theory of everything?** Theory of Everything (ToE) is the development of a theory which encompasses the whole gauntlet of what exists in our physical universe. This means bundling together a theory that handles everything right from micro (such as the tiny structure and origin of atoms), to the medium sized (solar system origin design), and the macro (galaxy design and origin). A ToE unites all these, bundles them together and connects the dots. Such a theory might explain our entire physical universe with everything connected.

Such a daunting task demands an extremely consolidated explanation. Can the Trillion Theory (TT) of this book make such an attempt? Yes, it feels it can.

Such an attempt at a ToE has been made throughout this entire book. New theories proposed here have attempted to uniquely tie together the full scope of our physical universe of atoms, matter, light, black holes, moons, planets, suns, galaxies, space and time.

For, included in any ToE is what we can see and what we cannot see. This is on both the micro and macro levels. So far, scientists haven't been able to see everything inside a tiny atom nor inside a naked black hole or a supermassive at the center of a galaxy. Nor have they seen deep inside of a sphere to see the black hole which is at the core of every sun, planet or moon. Trillion Theory states they are there.

TT conveys that the same laws and principles of recycling are seen throughout our entire physical cosmos. There is a direct relationship between spun matter in all the parts of our cosmos and on every scale of size right from the atom on the micro scale to spiral galaxies on the macro end.

In its simplistic model, TT offers a new paradigm view of our cosmos. Yet TT's explanations are simple.

TT alleges that scientific cleverness was behind the design of our cosmos. Pure genius is seen by deployment of just one basic material, namely light, to hold the necessary vast array of properties to construct the spheres of our cosmos. Then, also supply a method to recycle and perpetuate.

Trillion Theory shows similarities across our cosmos:
Similarities in type of spin:

▪ spin and pumping action are in an atom which can spin for billions of years with electrons orbiting a nucleus.

▪ spin and pumping action within naked black holes creates the force to spin light into matter.

▪ spin and pumping action are in every black hole which is at the center of a moon, planet and star in our universe.

▪ spin and pumping action are in supermassive black holes at the center of galaxies. Their pumping emits plumes from the poles of the black hole forcing light back out into space.

Similarities in formations of linear shape:

▪ linear is a formation shape seen right across our cosmos.

▪ solar systems are shaped like a flat disc with a central sun.

▪ spiral galaxies are shaped like a flat disc or a pancake with a bulge (supermassive black hole) at the hub.

Note: This flat shape is the result of the gravitational station sending out its gravity along a linear plane from its equator.

'Independent choice' for direction of axial spin occurs right across our entire cosmos:

To begin, some questions:

What direction do black holes spin on their axis?

What direction do moons/planets/suns spin on their axis?

What direction do solar systems revolve?

What direction do spiral galaxies revolve?

What direction does our entire physical cosmos revolve?

Are we a right-handed or a left-handed universe? Is cosmic spin direction always counterclockwise or always clockwise?

The answer is that this direction of spin is determined by choice across our entire cosmos. Namely, direction of spin is similarly the *choice* of the particular black hole at the center of any sphere, at the center of any solar system's sun, and at the hub of any spiral galaxy. This choice made by any particular black hole could be a clockwise direction or it could be counterclockwise. This directional choice is made at the birth moment of the black hole, or impacted upon a black hole after it has survived a Supernova, split into two black holes, spinning away from one another. Once this spin direction happens, that black hole will maintain that same spin direction for its lifetime, however long that lifetime is within the cycle it is presently living through.

Note: Astronomers have discovered that spiral galaxies in the northern and southern hemispheres of our skies spin either counterclockwise or counterclockwise. About 50-50.

So, we see right and left-handed spin across our cosmos of galaxies. Obviously, it's the supermassive black hole at the center bulge of a galaxy which sets the preference for the galaxy according to which direction it rotates on its axis.

This rotational spin is a commonality across our entire cosmos, prevalent with moons, planets, stars, solar systems, galaxies, and clusters of galaxies. It has yet to be determined which direction our entire linear cosmos spins.

This new innovative Trillion Theory model shows just how this commonality of rotation occurred right from the very beginnings of our cosmos up to the present, all-across our entire cosmos. Rotating black holes, busy spinning light into matter, created the spinning feature found in our cosmos.

Therein, the commonality across our cosmos is that black holes have produced the multitude of spheres seen in our universe. In commonality, every one of these spheres has a rotating black hole at its core running the show.

The direction of spin can be counterclockwise or it can be clockwise for different spheres and galaxies as dictated to them by their central black hole. In that way spinning black hole spheres are independent from other nearby spheres or galaxies as to direction of their axial spin.

Trillion Theory shows similarities across our cosmos.

As a Theory of Everything, Trillion Theory proposes that our universe known as RECYLIUN (Recycling Light Universe) has been recycling itself for a trillion years, the equivalent of 67 of the 15 billion year cycles of our cosmos. The material for recycling is light with all of its incredible properties, and the engine of spinning light into matter is black holes.

Light and black holes are common denominators found in every cosmic quadrant, and on every cosmic scale.

Trillion Theory discovers a new form of life.

'TT discovers brand new life form throughout our cosmos. To go where no man has gone before.'

We always thought the next form of life discovered would be aliens coming to our planet by starship or that our SETI (Search for Extraterrestrial Intelligence) would come back to us as radio waves from some distant planet. Instead, a new form resides at the cores of spheres; namely black holes.

TT shows how black holes are 'alive' - just a different type of alive than we have ever known. In TT, black holes have many life-type capabilities: they can spin prolifically; they can create a gravitational pull around themselves; they can exist at the core of a moon, planet or star for billions of years; they can hold matter around their bodies; they can hold other spheres in orbit; they can replicate; they subdivide their compartments in order to grow larger; they withstand a Supernovae by never being destroyed, surviving forever; and they can recycle light and matter of our cosmos.

So, a black hole eats, is greedy, spins light into complex atoms of matter, holds onto that matter for billions of years, grows in size, grows its tissue, replicates, adapts, changes strategy when needed, and has lofty goals. Seems very alive.

As the machine mechanisms of our cosmos, black holes have been masterfully constructed. One of the monumental tasks of future astronomers will be the finding of a way to look inside and examine the working parts of a black hole. (Trillion Theory has only made its best guess).

How extensive is a Theory of Everything?

Trillion Theory really provides an individual with a new realization of our physical universe. Each time we learn more about our universe we are amazed at just how complex it truly is. Yet, we are slowly puzzling that complexity into a mosaic which makes sense and helps us to fully understand.

For instance, a person gains a whole new perspective of how a sun amazingly cam to be or a planet/moon formed.

Isn't that the goal of a Theory of Everything, to provide us with a new realization paradigm?

Most spin, pumping action, and the gravitational forces within our cosmos go totally undetected by us. Stand and feel all the forces affecting your body at one time and you likely don't feel many of them. Yet, gravity is holding you to the surface of this planet. This planet is rotating on its axis so there is a movement of the air and atmosphere. This planet also travels in orbit around the sun of this solar system. Our sun travels, taking us along, within a spiral arm of our galaxy. Our galaxy spins past other galaxies as it travels space.

Trillion Theory shows magnificence - a GRAND DESIGN.

There are inadequate superlatives to ever describe the magnificence of our cosmos. If it were a contest, it's positive that our universe would win in a runaway for its simplistic yet uniquely scientific grand design.

Suppose that YOU were given the challenge of creating a universe and the rules stated, '*only one basic material and only one type of engine allowed.*' Such stringent rules would make the grand cosmic design extremely difficult. Of course, you would attempt to design a material and an engine as scientifically complex and versatile as possible. It would be wondrous to see how the inventions actually took place.

In actuality, light was designed with incredible science behind its properties, and then mass produced, as the mass material for our universe. Light, wondrous as a material and possessing so many unique properties that proper use of it affords a multitudinous variety of outcomes and products.

Light, this material which could be spun into over one hundred basic elements; able to withstand physical and chemical changes; able to exist in atoms of gas, liquid or solid and often move back and forth from one medium to the next; able to couple up as molecules to further the variety. Light, capable of forming the body of matter of all moons, planets, stars, and galaxies. Totally an indestructible material, light capable of recycling back and forth over a trillion year span, from light to matter – matter to light.

Black Holes were designed-constructed as the engines of our universe. Black holes are as unique as light, in their own way. 'If light can recycle over and over again into matter and always recycle back to light, then black holes are the engines which provide the power to make that recycling possible.' Black holes are such incredible engines. They spun and still continue to spin the moons, planets, stars, and galaxies of our cosmos. They also do recycle spheres into the next era. Black holes are gracefully powerful super fast engines.

TT displays how clever the design of our cosmos was in inventing and using only two basic items for that design. Both light and black holes are most worthy of being called super-duper scientific feats of the grand cosmic design.

Perhaps, the most amazing cosmic feature is the similar implementation of spin right from small atoms on the micro extreme to gigantic galaxies on the macro. All have similar principles neatly applied into their sculpting.

In this absolute cleverness we see strategic techniques which allowed our cosmos, once set in motion by light and black holes, to operate on a hands-off basis and requiring no further outside manipulation. TT says "that's incredible."

What to make of the inference of a GRAND DESIGN.

Humans are at an exciting point in existence. We have learned how to think more for ourselves with each new generation. However, science and church are still distant in cumulative attempts to figure out our universe.

But, before being labeled as a *creationist*, let it be known that TT advocates from a scientific perspective. Yet, TT does believe in an individual having a soul. Read the 2013 novel *The Trillionist* by this author for thoughts on that subject.

Yet, TT attributes everything to science. Even a person's soul - should it exist - is a grand scientific achievement by whoever or whatever did the inventing, design, and building.

TT claims our universe to be complex beyond belief. Why shouldn't it be? It has every right. Yet, its theories show the simplicity in materials and construction methods deployed in building. TT demonstrates an ultra-clever strategy behind cosmic design, with complexity hidden behind simplicity.

Do we need to include who and why in a ToE?

As an end goal, that is man's final determination – why are we in this universe? What is life about? TT does not at this stage aspire to resolve human creation. Rather, TT deals with a physical universe. But stating how clever and strategic certain cosmic events are in our cosmos is not creationism, rather it is TT pointing out grand scientific methods.

Think about it! Every time we move to a new level of cosmic understanding, we uncover greater complexity. The discovery of atoms took us from simple to ultra complex. TT contends that the uncovering of the ultra complexity of our cosmos will continue to escalate. As we discover more about our universe, we will find more complexity at each new level.

Further on the complexity notion, TT contends that any thought of a divine omnipotent creator behind the wheel will also move to a much higher ultra complex rung, but only after we can fully understand our physical universe. Whether we use terms such as creator, artisan, architect, corporation, or computer hologram program, all this is bound to become more ultra complex, following a similar path which we travel in gaining a greater understanding of our cosmos. TT shows our cosmos as being a finely oiled machine, purposefully operating and scientifically designed.

Yet, TT doesn't endear *creationism* which advocates that all things were created as they now exist. With that TT can't agree. TT takes evolution further, proposing that our cosmos evolved and grew in size over a trillion years of history.

Creationists believe everything was created from nothing as described in the bible. Instead, TT shows how all matter in our cosmos formed from huge amounts of an energy source which was and still is in ample supply.

TT submits how something, a trillion years ago, was obviously designing our cosmos in a pre-planned clever strategic recycling manner. This growth has occurred on cue from cosmic rules which were cleverly implanted pertaining to the formation and the growth of spheres and galaxies.

Trillion Theory demonstrates just how strategic cleverness was used to build our ultra complex magnificent universe. As a marvel beyond any comparison, our universe is a living growing entity incredibly perpetuating from one 15 billion year cycle to the next, seemingly going on forever, and all according to a very strategic system.

The question arises: Is our universe mere happenchance, purely an accident? Or, did the hand of some creator-artisan or architect place scientific design onto its construction?

When you read TT, your thoughts as to a creator-type-entity might change, and whether TT influences this is not its direct intent. From a TT perspective, it wants to sit on the fence, keeping an open mind. Our complex universe may yet have a totally different design component which we are still years from discovering – hopefully not light years.

Final thoughts on a Theory of Everything (ToE).

For humans to first solve our physical universe and then go beyond that, it is critical to that effort to have in-hand a true Theory of Everything. It must explain a physical universe from the micro world of subatomic atoms right to the macro world of gigantic galaxies and also encompass the entirety of our cosmos to its outer edges and beyond.

This has never been accomplished before.

Trillion Theory states there are changes we can make:

- Quit thinking of our universe in only rudimentary ways.
- Don't be fooled by what your eyes see - dig deeper.
- Require real proof. Quit trying to fit into Big Bang.
- Open up 'think tanks' and brainstorm about our cosmos.
- Think more like creators: how would I build a cosmos?
- Work to put the genie into the bottle, like a creator.
- Envision fantastic science as the builder of our cosmos.
- Think strategically as to how our cosmos was built.
- Go out on a limb, that's where the fruit is.
- Think big – our cosmos is complex beyond belief.
- Look at radically new ideas such as Trillion Theory.

Interviewer: "Does TT offer an explanation strong enough to be a new paradigm to replace the Big Bang?"

Author: "You mean does TT have a snowball's chance in hell? I'm not sure. A paradigm shift occurs when new theory replaces old thought and the new idea becomes accepted norm. TT certainly offers a whole new paradigm."

Interviewer: "An astronomy magazine says that our cosmos is supposedly homogenous and isotropic? However, TT uses the term linear. So does TT contend that our cosmos is flat?"

Author: "Bang and Inflation theory both incorrectly assume that their supposed explosion starting our cosmos expanded matter out from the center evenly to every point in the cosmos. If Big Bang theory were correct, our cosmos should be a perfect roundish shape, homogenous (uniform) and isotropic (identical) in every directions from the center.

But, it isn't. More linear is the new universe shape. TT affirms that while the entirety of cosmos is quite linear, it's not pancake flat, but rather the shape is more like a fat wide cucumber. Early in cosmic history a trillion years ago, the first black holes ate from the ocean of light along an extension their equators along a linear plane. But over time, as more black holes entered the fray, they axially tilted to more strategic angles from the linear plane when battling for optimal positions against other black holes. Therein, our cosmos might be fairly linear overall, but not perfectly flat, more the shape of a double-wide cucumber."

CHAPTER 17
A NEED FOR REAL PROOF

Trillion Theory says, "Big Bang needs far better proof than, 'Galaxies seen through telescopes are receding from one another, meaning there must have been a big explosion creating our cosmos.' That's insufficient proof."

Trillion Theory flatly states that current widely accepted proofs of the Big Bang are not sufficient proofs. Collected scientific evidence has been twisted in an effort to make everyone accept Big Bang as the origin of our universe.

Also, it is not sufficient for astronomers to measure the age of older stars at 13.7 billion years and conclude that's how old our universe is. Suppose someone saw the oldest trees in a forest at 200 years, should they simply conclude that the forest is 200 years old, when in reality it is millions? The difference is that our cosmos is the ultimate recycler, totally capable of hiding the past cycles of our cosmos.

There are 3 things to accomplish here:

♦ Show that Big Bang theory is a false unproven theory.

♦ Show how Trillion Theory is a possible correct theory and how it tackles the claims made by other theories.

♦ Provide possible proofs for Trillion Theory, and suggest ways that astronomers and astrophysicists can assist.

Say 'no' to Big Bang hoax. Get rid of Big Bang.

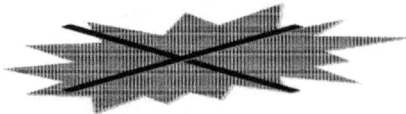

If Big Bang is an incorrect kaput theory, it belongs right alongside the other defunct false beliefs, namely the 'flat Earth' and 'Earth as center of our universe.'

However, Big Bang cosmology is still promoted today as an established fact and taught in schools and universities. That theory is strongly ingrained in millions of people. So it must be right? But, over the past 60 years, discoveries have often been twisted so that they'd fit neatly into Bang.

But, the Bang has everything wrong as to the origin, age, and the formation of the spheres of our universe. It claims that our universe began at a source point which exploded, expanding across space some 13.7 billion years ago. Since Big Bang had no way to show how spheres, solar systems, and galaxies formed, Nebular Theory was recruited to state how gas clouds supposedly whirled, eddying into spheres.

But, Trillion Theory adamantly says 'no' to Big Bang. Yet, it is necessary here to examine Bang's claims since nearly every reader has been taught to accept the Bang as law. So, before there can be a new sheriff of universe theory, either Bang has to be proven wrong, or a new theory proven right.

Paradigm shift to a revolutionary new cosmology model only occurs when one new theory replaces old thought and the new theory brings some paramount scientific proof.

Declaration: Big Bang's simplistic depiction of the origin of our cosmos is a false and outdated incorrect theory. It is based too heavily on happenchance dreadfully failing any test to properly explain many complexities of our universe. Big Bang simply leaves such subjects to one's imagination. Note: Renowned pioneer Radio Astronomer Grote Reber had always been skeptical, stating that Big Bang was bunk.

Why Big Bang theory refuses to go away, even though there are many huge gaps in Big Bang theory.

There is difficulty in challenging an entrenched paradigm such as a Big Bang. For the past 60 years, astronomers and astrophysicists have tried to find monumental proofs for Big Bang: redshift readings of galaxies ebbing away; microwave background radiation indicating a hot cosmic glow; and gravitational waves indicating past expansion. But, all these supposed Bang proofs are inadequate says Trillion Theory.

So, in today's modernistic world, it shouldn't be difficult to know when an old theory is *blowing smoke*. It should be relatively easy to replace an old incorrect Big Bang theory?

Problems had already begun to surface for Big Bang

A few years ago, astronomers made the discovery of a star which they calculated as 18 billion years old. In 2013, this paradox continued unsolved with the discovery of the impossible Methuselah Star, at 16 billion years ancient. If the origin of our cosmos occurred 13.7 billion years ago, how could two stars trump the age calculation of Big Bang? This is definitely a paradox and a fly in Big Bang's ointment.

Big Bang relies on false premises, Trillion Theory says:

• **False to Big Bang** for estimating the age of our cosmos at only 13.7 billion years by incorrectly using our current sky as the basis for this age calculation.

• **False to Big Bang theory** depicting cosmic origin from a central explosion of matter outwards.

• **False to a Big Bang** conclusion that galaxies receding from each other proved an explosive origin to our cosmos.

• **False to Nebular Theory** of how planets, moons, and stars supposedly formed from nebulas of swirling gas clouds.

TT attacks Big Bang's three main supposed proofs:
TT refutes Big Bang's supposed first proof: Redshift galaxy readings indicate that galaxies are receding away from each other is interpreted as Big Bang's proof of an explosion origin. Bang Theory first came about as a theory because of this redshift discovery. Simplest conclusion at that time was that a central explosion sent all cosmic matter outwards.

But, TT contends that our cosmos is pulled outwards by an ocean of light which exists on the perimeter surrounding our cosmos. Galaxies are continually attracted towards this perimeter, thereby expanding space's outer boundaries.

TT argues against Bang's supposed 2nd proof: Background radiation supposedly indicates that our cosmos began hot leaving behind a microwave glow. TT offers other reasons for this radiation, affirming that microwave radiation should be everywhere. TT states that in our earlier cosmic cycles, small galaxies exploded when their massive sun went Supernova, emitting tons of radiation. However, that is much rarer today as supermassive black holes have evolved to keep galaxies intact, and supermassives live for 100's of billions of years.

TT discredits Big Bang's supposed 3rd proof: Gravitational waves might indicate past expansion right after a supposed Big Bang. Recent discoveries showed ripples in time-space which the top astrophysicists think are the very first tremors (aftershocks) 380,000 years after the Big Bang.

TT discredits this, arguing that these ripples are leftovers from a trillion year history which witnessed small galaxies totally exploding when their massive sun went Supernova and exploded an entire galaxy. These were massive recycling explosions, leaving behind time-ripples in space.

TT declares, there never was a Big Bang cosmic start!

TT declares that Big Bang never happened. Trillion Theory demonstrates how solar systems and galaxies are structured to the nth degree. Organizational laws repeat themselves in all solar systems during a trillion years of cosmic history.

TT contends that Bang is a fallacy, built upon pillars of sand. When future generations realize true cosmic origins, they'll look back at Bangers and laugh hard, just like we view those of ignorance who once believed in a flat Earth.

Today, lack of a broader understanding of our cosmos is holding humans back in a multitude of ways. Big Bang has only sufficed by being a convenient explanation. But, there exists an absolute need for the next generation of scientific discoveries to uncover more universe secrets.

TT attacks Nebular Theory too.

Nebular attempts to assist Bang in addressing how solar systems formed. In Nebular, supposedly gas clouds from Bang coalesced. Stephen Hawking states that stardust of swirling nebula clouds gravitated and assembled all the gas particles into spheres in our solar system. TT responds that Nebular is as ridiculous as the Bang. Neither explains the radically varying axial tilts of the planets in our solar system.

Also, after the supposed Bang occurred, how come stars don't contract in the extreme coldness of space, but rather expand? Ten billion years is an extremely long cold time.

How come volcanic flows on planets still fire outwards, when a planet's core should actually be freezing in space's cold depths? How come spheres don't simply freeze solid? Bang supporters say, compression of the surface started the fire in the pit of planet Earth. Trillion Theory won't buy that.

The real question should be: what internal force is really causing the interior of that planet to heat up and get hotter?

In a recent study, scientists are coming closer to shedding light on what actually causes volcanic eruptions. TT applauds this as a step in the right direction, as TT contends that cores in cosmic spheres all heat up more over time as the earliest spun atoms begin to unravel at the spheres core. Therein, hot atoms of lava search for a path to a sphere's surface.

Using a supercomputer, these scientists saw new images coming from well below the Earth's crust. They found rivers of lava which were fed by two 'superblobs' of gigantic size. Thus, magma rises to the surface via columns through fissures from these hot deep molten rivers.

This discovery does to show that TT is correct in depicting how our sun, billions of years ago, turned from being simply a massive planetoid into a full scale heat emitting sun.

A planet sitting in the dark extreme coldness of space. What really started the fire deep in the bowels of its core? What expanding force is sending pressurized lava to the surface?

Nebular Theory **incorrectly** premises our solar system as having begun some 4.5 billion years ago as a large irregular cloud composed of gas and dust. But, Nebular does nothing to explain where elements found on Earth came from.

Nebular theory **incorrectly** proposes that a gravitational collapse pulled a gas cloud close together to form our solar system, forming collection spherical clumps. Nebular has no way of accounting for where gravity actually came from.

Trillion Theory attacks Inflationary Theory.

Inflation Theory by Alan Guth tried to answer a classic Big Bang conundrum: why does the cosmos appear flat? Big Bang's explosion should have been isotropic (outwards in all directions). TT purports our cosmos as linear, not roundish. TT offers a totally different reason for this linear than the supposed inflation explosion put forth by Guth.

Trillion Theory takes on Stephen Hawking.

Even Stephen Hawking, the often admired and renowned English theoretical physicist and cosmologist, in April 2013 admitted to a rather large blunder. Till then, he'd thought that light swallowed up by a black hole was lost forever. But, Hawking recanted his stance admitting that radiation does in fact escape from a black hole. This agrees with TT showing light escaping our sun after billions of years of entrapment, finally eluding the clutches of a black hole at the sun's core.

Hawking has stated that light can never escape from a black hole. But, TT describes how light always eventually escapes a black hole, even if it takes 10's of billions of years.

Show Trillion Theory (TT) as the correct theory. All other theories incorrectly show a 13.7 billion year cosmos.

TT declares the incorrect 13.7 billion year age estimate is a result of other theories focusing solely on the age of the stars in our present sky. TT digs deeper finding that the stars have recycled many times over in the past a trillion years.

Why proof of Trillion Theory (TT) is vital?

Because TT covers so many cosmic aspects, there will be numerous ways to prove Trillion Theory. TT states that black holes over the past trillion years of cosmic history are the builders, the organizers, and the operators of our cosmos.

Stated rather directly, 'Trillion Theory advocates that black holes are the builders of spheres, operators of solar systems, and the organizers of galaxies in our cosmos over the past trillion years of cosmic history.'

Proof of TT is vital for man's next steps into space. But for now, TT is mere conjecture - great fodder for all discussions.

So give TT a chance. TT deploys a more advanced paradigm model depicting a totally different way our cosmos began, grew and recycled over a trillion years. And hopefully new scientific methods will provide proofs. For at the outset, a theory is just a proposed hypothesis stating a definitive belief about a phenomenon in our cosmos. When a theory is proven, it becomes accepted norm, the new PARADIGM.

Proofs for radical new Trillion Theory (TT).

Always, science is called upon as the means to prove or disprove a conjecture theory. It is one thing to say it is so; it is quite another project to prove or disprove a new theory.

The biggest obstacle to proving theory is the short time of man's existence compared to a trillion year cosmic history. Man's time goes by in a blink of an eye, while our cosmos recycles at a snail's pace. As actor George Clooney, of the movie *Gravity*, said about man's existence compared to the cosmos: "I'm very aware of just how brief life is."

For instance, to catch the action at the start and end of one solar system we'd need to run a video for 15 billion years. So instead, we are relegated to piecing together the snail-like action from hundreds of different solar systems in various stages of their cycles. The good news is we humans are very resourceful. As Carl Sagan remarked, "Extraordinary claims require extraordinary evidence."

That was the case in November 2013 when American Press announced, 'Earth-like planets fill our galaxy.' A study, published in the National Academy of Science, found our Milky Way is teeming with billions of planets circling stars similar to our sun. Trillion Theory had already clearly stated that solar systems are the norm throughout the cosmos, and a star without a solar system is considered an abnormality.

Possible proofs to authenticate new Trillion Theory (TT).
Proof Method 1 for Trillion Theory (TT).
(Find black holes in a Supernova-Obliteration graveyard).

Doable. Powerful telescopes may someday peer deeper into landscapes after a Supernova. TT contends that there will be many naked black holes (remnants from the old destroyed spheres of the old solar system) occupying the area and filling up with light to begin the next cycle.

Powerful space telescopes have difficulty seeing into the brightness of a Supernova and the Obliteration of a solar system because of the huge flash of light. Just as hard to see is the dark blackish holes left behind right after Supernova and Obliteration. TT says that survival black holes dominate the graveyard after the death of a solar system.

If many black holes are found in the aftermath death of a solar system, then that proves Trillion Theory. TT details how the black holes have reproduced and are now preparing to rebuild that solar system into the next 15 billion year cycle. However, that graveyard might also be very bright instead of pitch black, as the black holes in that area will be pulling in streams of light and spinning them into matter around their bodies. The black holes may be hidden behind veils of light.

To find this proof, an astronomer would really have to be Johnny-on-the-spot, for being late may mean missing those naked black holes which filled up again to appear as new moons and planets. Too late and each black hole would be hidden away once again within a sphere's core.

Proof Method 2 for Trillion Theory (TT).

(Axial spin direction varies with solar system spheres).

Nebular Theory incorrectly states that every sun, planet and moon formed by a certain nebula should have the same direction to its axial spin. But, they don't.

If the nebular cloud supposedly spun coalescing to form our sun and planets, then that spin should have given all the planets the same direction of spin on their axis, but it didn't. Nebular can't explain why certain spheres rotate clockwise on their axis while others rotate counterclockwise. Example: Earth spins counterclockwise on its axis; Venus clockwise.

So, let's investigate bizarre Venus. It is most likely a mirror sister of Earth. When, the past solar system occuping this area was obliterated by its old sun going Supernova, one of the average sized black holes of a planet survived and split its black hole into near twins: Earth (7,926 miles in diameter); Venus (7,521). Venus and Earth are a little different in size simply because their black holes ate and spun light into matter at different rates. Fierce splitting spun their two black holes away from each other, applying two directly opposite rotational directions: planet Earth spun counterclockwise, while planet Venus rotates almost perfectly clockwise.

This discrepancy is explained in TT. After a Supernova within a solar system, the split of the surviving black holes can apply either spin direction to the two new black holes.

Proof Method 3 for Trillion Theory (TT).
(Axial tilts of solar system planets differ).

Nebular Theory incorrectly states that every planet which was formed by a certain nebula should have the same axial tilt when riding in orbit. But, they don't.

Those varying axial tilts of the planets in our solar system should not be there if there had been a Big Bang followed by Nebular. If the nebular cloud supposedly spun coalescing to form spheres, then that spin should have given all planets similar axial tilts, but it didn't. Nebular can't explain why.

Astronomers have shown that in our solar system the tilts of planets differ, and nebular theory has no way to explain. TT shows our solar system's planets as operating under their own tilt. Every sphere in our solar system has an axial tilt unique to itself. This could be proof of Trillion Theory.

TT shows how each black hole forming a planet spun at different axial tilts right after the old solar system which previously occupied this spot was destroyed. Varying tilts for each new black hole occur, as each black hole will search for its best strategic angle for attracting light.

So now, each planet sits in orbit in an axial tilt which was decided by its black hole back when it was trying to find its optimal angle to best fight for light. This tilt has remained with the sphere even after it found its orbit around our sun. Example: Earth's axial tilt is 23 degrees.

Now swing over to weirdo Uranus, which spins on its side because TT says all black holes can independently determine their own axial tilt when they are battling for light to spin into matter. Uranus is direct disproof of Nebular Theory and Big Bang. Uranus axial tilt is direct proof of Trillion Theory.

Proof Method 4 for Trillion Theory (TT).
(Find the black hole at the center of a sphere).

If we were to dig to the core of any moon, planet or sun, TT proclaims we would encounter a spinning black hole. This would be direct proof of Trillion Theory.

However, the black holes which we would find at the core of spheres would appear much different than any distant naked black hole astronomers have perceived.

With spheres, it's difficult to differentiate the black hole at the core from the rest of its outer body, since they blend as one. The black hole at the core gives the sphere its direction of spin on its axis, and holds surface matter, plus smaller spheres in orbit. It'd be an ultimate rush to travel to the inner depths of a sphere and dig deep into its black hole.

Proof Method 5 for Trillion Theory (TT).
(Spin light into matter in a physics laboratory).

The 5th proof of TT may be possible right here on Earth, and prove TT beyond a shadow of a doubt.

In my futuristic sci-fi novel *The Trillionist,* the young lad in the novel invented a machine called a Quantronix which artificially did the task normally assigned to black holes. Quantronix was a large fat machine with a long giraffe-type neck sticking out the top of a domed building. Quantronix attracted light spinning it into atoms of rare elements.

The young lad succeeded at becoming our universe's first inhabitant to create matter from light. Perhaps in the future, humans will invent a Q-machine, to become artisans. Such a discovery would open up numerous hidden secrets about our universe. Such achievement would prove Trillion Theory, furnishing proof of how black holes formed our cosmos.

How modern day astronomers and astrophysicists can assist with the finding of proofs for Trillion Theory.

In the future, astronomers and astrophysicists will have a plethora of opportunities to assist in proving Trillion Theory.

Astronomers are learning more about the order brought to a galaxy by the supermassive black hole at its hub. But, they struggle to find rhyme or reason to smaller black holes. They ask, 'why the two types?' Without being able to answer.

It's incorrectly thought by those peering through huge telescopes that black holes are only formed as the remains of a sun gone Supernova, as it explodes and then implodes. It wasn't rationalized till TT that the black hole had initially formed the star around itself, billions of years earlier.

Today, gladly for TT, astronomers are eagerly studying black holes. Their discoveries will immensely assist TT to gain acceptance as a theory which better explain cosmic origin.

Assistance to TT also occurs elsewhere In an Associated Press article dated October 28/2015, researchers who were studying comet 67P/Churyumov-Gerasimenko did note significant oxygen (O2) presence emitted from the back of the comet. The University of Michigan scientists Kathrin Altwegg and Andre Bieler stated that this discovery could challenge theories of how our solar system was formed. theories presuming all matter was heated and then cooled. TT contends that cosmos didn't begin hot and then cooled.

TT theory declares, "Someday astronomers may build a telescope powerful enough to allow man to climb high enough to peer over the fence and see into the back yard of the Artisan who scientifically designed light (as the material) and black holes (as the engine-like builders) of our cosmos."

Interviewer: "Can a future bigger telescope than the Hubble Space Telescope help us to learn more about our cosmos?"

Author: "Of course. Even today, the Hubble Space Telescope is man's greatest friend, as new observations of the distant cosmos offer new looks. Interpreting those sites in the correct manner has and still is a difficult daunting task. TT's opinion is that Big Bang Theory is holding man back. So, the time has come to reopen minds. 1949 till now has been too long for an unproven Big Bang to be the cornerstone theory of our cosmos. But, it's never easy to kick out a squatter."

Interviewer: "TT proclaims that astronomers can misinterpret their discoveries, trying too hard to prove Big Bang."

Author: "Absolutely possible. You can't place a square peg through a round hole. Often new discoveries get wrongly skewed and interpreted to fit nicely into Big Bang."

Interviewer: "But, these theories which you are slamming are firmly entrenched. Big Bang has world-wide acceptance in scientific journals, magazines, books, novels, universities, school rooms and the television programs of our world."

Author: "Hold it. World-wide acceptance is not proof of a Big Bang. Don't you recall back in history that nearly everyone incorrectly presumed Earth was flat? That later proved to be a fallacy. Often incorrect theories become a paradigm simply because of their source. TT contends that Big Bang is such a mistake – a sand castle ready to be washed away."

Interviewer: "TT is seen even taking-on renowned Stephen Hawking. Maybe TT should take on someone less revered."

Author: "Good point. But, I wouldn't take on or disagree with a Copernicus or a Newton, because their beliefs are proven."

CHAPTER 18
FINAL RECAP AND WINDUP

Trillion Theory (TT) makes the claim: A universe is no accident, nor pure happenchance, but rather the result of the deployment of the pinnacle in scientific knowledge and inventiveness, black holes being prime examples - the true purpose of which TT endeavors to uncover.

Summarize the Goals of Trillion Theory (TT):
Main Mission of Trillion Theory – a new theory. The mission of TT is to propose a plausible new theory depicting how our cosmos originated one trillion years ago, then grew to its present prodigious size. And focus on the major role played by black holes in that growth. This mission seeks to radically jump-start a game-changing view of our universe into a new Trillion Theory paradigm.

Second Goal of Trillion Theory – Proof.

TT offers 5 possible ways to PROVE its theories using future scientific means. Such proof would allow new Trillion Theory to become the NEW PARADIGM model for properly understanding and explaining our cosmos.

Third Goal – TT says '*No*' to Big Bang.

Declaration: Big Bang never happened 13.7 billion years ago. It's a 60 year old theory, hanging on by its fingertips; a sacred cow waiting to be tipped over; offering no proofs; too simplistic to explain our cosmos. In reality, Bang Theory only came about because of the discovery that showed cosmic expansion. At that time, the simplest conclusion was that a central explosion had sent all matter outwards. However, Trillion Theory suggests better reasons for this expansion.

Short Summary of Trillion Theory in this book:

• Trillion Theory sees an ordered system – not chaos.

• Trillion Theory sees 'orderly arrangement' in our cosmos resulting from cosmic laws of engagement between spheres. Laws seen throughout billions of solar systems and galaxies.

• Trillion Theory asserts that our cosmos began small a trillion years ago and then grew to its present gigantic size.

• Trillion Theory expounds that the history of our cosmos has been one 15 billion year cycle after another; each new cycle intertwining with the last; growing grander with each cycle to our present 67th cycle hosting 73 quintillion stars.

• Our universe's history has seen 67 cycles during its trillion years, with each of these cycles spanning 15 billion years.

• During a cycle, stars age out by going Supernova thereby recycling themselves and their solar systems into new ones.

• The spheres and solar systems in our present sky are just the latest rendition to occupy our current 15 billion year segment cycle of cosmic history.

• Spheres and solar systems recycle about every 15 billion years, while the galaxies in which they reside can be much older because of the supermassive black hole at the core.

• Trillion Theory suggests that our cosmos originated from an endless ocean of static (frozen-like) light.

• Trillion Theory shows how the endless ocean of light has over history, and still is, the supplier of the light being used as the material to make the matter of our cosmos.

• Trillion Theory states that the billions of spheres, solar systems, and galaxies all originated from this ocean of light.

• Trillion Theory describes an outside force pulling contents of our cosmos ever outwards towards the ocean of light.

- Light spins to form atoms (elements) of universe matter.
- Light carries in its tool box all the properties for matter.
- Light locked away as an atom of matter will spin for eons.
- Light will always want and succeed in someday escaping its atom and depart away at the speed of light.
- Naked round black holes are the power engines designed with spin speed methodology to spin light into matter.
- Trillion Theory shows how black holes spin forever, on and on, it's what they do. And on the micro scale, they also spin the atoms (elements of matter) during their sphere building.
- Naked black holes spin matter to fill their belles and to form planets, moons and suns around their spheroid shape.
- Trillion Theory shows black holes as the builders of all the spheres of our cosmos. They are also the organizers and controllers of billions of solar systems and galaxies.
- Therefore, Trillion Theory advocates that a black hole exists at the center core of every planet, moon and sun (star).
- XL planets have the possibility of evolving into a sun. But, most never get the chance.
- Neither light nor a black hole can ever be destroyed. They are the two main indestructibles of our recycling cosmos.
- Black holes replicate, splitting from one to two when a sun goes Supernova and obliterates the planets and moons of its system. The black hole at the core of each sphere survives.
- Solar systems grow larger each recycle to where they can hold hundreds of planets/moons in orbit around a sun(s).
- One solar system can be broken apart during a Supernova; thereby two new similar sized solar systems can result.
- Size matters, as larger black holes overpower smaller black holes to gain control of solar systems as their suns.

• A galaxy becomes the hotel island oasis of the cosmos. They provide the template holding many solar systems in place around a hub. Within the galaxy, solar systems recycle to thereby increase the number of spheres in that galaxy.

• As solar systems increase in number in a certain area, the most powerful supermassive black hole of that area takes over their hub forming a galaxy. That supermassive black hole can survive for hundreds of billions of years.

• Spin, provided by the super spin of black holes, is the main motion imparted to our cosmos. This spin is seen in black holes, planets, moons, suns, stars, galaxies and even atoms.

• Linear is the main configuration between the spheroids of our cosmos. This linear configuration is seen with the rings around planet Saturn, with the pancake shape of our solar system, with the Frisbee appearance of spiral galaxies, and with the elongation of our entire cosmos on a linear plane.

• Space is a void empty byproduct left behind when huge amounts of light depart an area. This departed light is spun into matter around a black hole forming a sphere.

• Time is imported into any area where light is slowed and spun into matter. Time is the duration measurement during which light stays spun as atoms round a black hole.

• Trillion Theory proclaims our cosmos was 'placed on the clock' (imported time) when black holes first spun light into matter to form the first early spheres of our cosmos.

Conclusions made by Trillion Theory (TT):

• TT suggests that the building and recycling set-up of our cosmos can be likened to a large project. The assignment was to build a complex physical universe while utilizing only one ultra-versatile indestructible material, namely light.

• Light carries in its tool box a myriad of great properties.

• Black holes are the engines of the cosmos which can spin light into a multitude of elements.

• The next step of the design project was the establishment of definitive cosmic laws of engagement between light and black holes, also between black holes in combat, and also between the spheres they build in co-operative situations.

• Trillion Theory suggests that our cosmos was a most highly planned project, utilizing absolute masterful technology and science for the design and the methods of operation.

• Trillion Theory suggests that a similar deployment of top level science was placed into the growth and recycle module for our cosmos, to such an extent that the recycling process would be able to function of its own accord onto perpetuity.

• Trillion Theory suggests that our cosmos is a technological marvel involving: scientific inventiveness, strategic purpose; unique clever design; and organizational cosmic laws for black holes, spheres, solar systems, galaxies, and recycling.

• Trillion Theory suggests that the designers of our cosmos left several clues to help inhabitants decipher: a rainbow shows a tiny sample of light's unique properties; lightning releases its powerful light and heat; an atomic explosion releases the power of the atom showing what is inside; a Supernova releases the huge power of an exploding star.

If Trillion Theory is true, there are huge implications:

• The implications of a trillion year old cosmos could be dramatic for humans both on a physical and spiritual plane.

• If there have been living beings on planets during any or all of the other 66 previous 15 billion year cycles, then life may have abounded during a trillion years of history.

• If every star has a solar system, as Trillion Theory suggests, this greatly ups the possibility of life on many planets which have a Goldilocks situations of not too hot, nor to cold.

• Some past civilizations may have reached incredible levels of technology, hundreds of billions of years ago.

• Civilizations which reached incredible levels of technology obviously discovered that their solar system would someday recycle; therefore they would have had to settle colonies outside of their own solar system in order to increase the odds of the perpetuation of their society.

• And for those who believe in reincarnation, Trillion Theory takes the possibilities to new heights. Read my novel 'The Trillionist' to discover that reincarnation has a deep and rich history over the trillion years of our universe. Reincarnation might have the propensity to be a gigantic feature, as any spirit may have reincarnated thousands of times.

Who helped the author develop Trillion Theory.

No one. Trillion Theory is solely this author's doing.

Author's Disclaimer: Theories in this book were developed solely by the author Ed Lukowich. These Trillion Theories are not based upon any other person's theories which may somehow find any similar basis. Therein, no other person may claim to be the originator of any of these Trillion Theories.

I've freelanced unencumbered to write my own universe theory. I've always interpreted new info about our cosmos in my own way. Many of my new ideas came from my strategic thought processes as to how a universe might optimally be constructed. I gave new purpose to black holes and how they operate to build and grow our cosmos. With light, I envisioned it spinning to form atoms of matter.

However, I've had many heroes of history to look up to: Copernicus, Galileo, Newton, and Einstein; and then also Carl Sagan who had this wonderful quote, 'It is much better to grasp the universe as it really is than to continually persist in delusion, however satisfying and reassuring.' TT points this quote directly at Big Bang. We need a new paradigm.

Yet, it pleases me to live in a world where it's possible to gain insight into explaining our cosmos. Over history, truth has always emerged victorious, even if taking decades or centuries. Inside me, there had always been an innate need to better understand life. Like many people, my eyes have looked to the sky, wondering what this world is all about.

Today, I've taught myself to see our night-time sky much differently. Today, I see spheres made of incredibly tightly spun light-to-matter. Then there is the endless empty space left behind as all that light departed into the spheres.

The original ocean of light, from a trillion years ago, is now gone from our seeable sky. Instead, in our viewable sky, all light (tons and tons of light) is presently tightly packed as matter inside of the sky spheres. Space has replaced the area previously occupied by the static ocean of light. These huge areas of space are now just the weightless, dark, cold, void byproduct left behind when the black holes spun all the occupying light into tight tiny atoms of hard matter. The light which spun into the spheres took with it all the weight and left space behind as a voided emptiness. Clever, as light and black holes erected space as a cosmic shell to hold spheres, solar systems, and galaxies, and be used in the future as a highway to travel.

Mastermind physicist Albert Einstein's work helped me tons. His formula $E=mc^2$ (or $mc^2=E$) showed unbelievable amounts of energy stored within an atom. Energy and matter had this back and forth forever relationship. Matter and energy recycle back and forth, never being destroyed.

That formula is now ingrained into Trillion Theory. So when Supernovae occur, matter escapes back to light trying to depart the scene, and when a naked black hole captures and spins that light, it returns to being matter.

To make my point, imagine we blow up Jupiter. Planet Jupiter is 88,736 miles in diameter, that's a lot of mass. But, here's the key, every atom of Jupiter will simultaneously split and explode. We know of the staggering release of energy from a single atomic explosion; m (mass) multiplied by c (light speed 186,282 squared) = E (967,000,000,000,000,000). The pop that genie releases is tremendous pent-up energy.

Quick ballpark math: energy (light, heat, radiation) from Jupiter atoms exploding takes up 1 million X 88,736 miles as the diameter of Jupiter = 88 billion miles. The release of all the atoms of Jupiter takes up far more area than the 7.5 billion miles diameter that our entire solar system occupies.

The main point is that a moon, planet, or star is more than just a hunk of matter. TT says, 'What comes out is what originally went in.' The amount which comes out is jumbo.

Humans figured how to split an atom, but we have yet to discover building an atom. TT admits, 'Nothing worthwhile is easy, least of all building an atom.' Carl Sagan stated, 'No great idea ever started without first being thought of as ridiculous.' TT hopes that, 'Someday we may climb high enough to peak over the fence into the Artisan's backyard.'

Defending Trillion Theory (TT).

The author of Trillion Theory has the newest cosmology theory available to depict the origin and operations of our cosmos. At this early stage, TT has a strong need to defend its efforts from its critics and also a further need to acquire substantial support in terms of building a large following.

TT's ideas are huger than huge. But, for the moment they may be a little bit sci-fi, as its concepts are much different. However, TT does have a propensity to someday be proven.

At this stage, no matter who the expert is, they could dispute TT, but they cannot prove TT to be wrong. In realistic terms, TT let the genie out of the bottle, letting even more of the secrets of our cosmos escape out to us.

Trillion Theory is hopefully destined to earn more support for its new unique ideas. Astrophysicists and astronomers are continually discovering more things out about the black holes of our cosmos all inline to support Trillion Theory.

Final author comments:

Thank you for reading this book. My ultimate hope is that you now think 'Trillion' and 'Black holes,' as I do. Or, some corridors have opened for you to new cosmic ideas.

No doubt, my new Trillion Theory will be bashed and trashed by many. And that's good. My philosophy is that the person who opposes can also help a project. My passion outweighs my fears. For me, a risk well worth taking.

My intent is to have no writer's regret. For, TT is either right or wrong, accepted or rejected on its own merit. In the long run, new scientific discovery will provide the proofs. For that exact reason, it was a mistake by all those who placed a totally unproven Big Bang on too high of a pedestal.

Here are some quite recent dramatic examples where man has had moments of being wrong: Western Union 1876 memo read, 'This telephone has too many shortcomings to be seriously considered as a means of communication;' Lord Kelvin 1899, 'Heavier than air flying machines, impossible;' Charles Duell 1899, U.S. Patents Office, 'Everything that can be invented has been invented;' Robert Millikan, Nobel Prize for Physics 1923, 'There is no likelihood man can ever tap the power of an atom;' Thomas Watson 1943 chairman of IBM, 'There's a world market for maybe five computers;' Ken Olson 1977, founder of Digital Equip, 'There's no reason anyone would want a computer at home.'

And more often than not, it takes time to install a new paradigm theory. It was 150 years from the time Copernicus first whispered that the Earth was not at the center of our cosmos to the acceptance of his new theory which placed our sun at the center of our solar system.

Interviewer: "Ed, you seem to be on quite a mission."

Author: "My mission with my new Trillion Theory is to really jump-start a new radically organized view of our cosmos on a going forward basis. I hope and believe that my new theories contribute in some way to being a game changer."

Interviewer: "Is your goal to establish a new paradigm?"

Author: "Definitely so. My **PRIMARY GOAL** is to propose TT as a sounder more plausible theory, a new paradigm model to explain how our cosmos began and then grew and evolved to its present incredible size. The **SECOND GOAL** is to suggest and find methods whereby Trillion Theory could be proven in the future. The **THIRD GOAL** is to rid ourselves of the old view; the stuck-in-the-mud Big Bang."

Interviewer: "Ed, obviously you've put plenty of energy and thought into your TT cosmic model. Okay, I ask again, tell me more of how your theories came to your mind."

Author: "I'm not a scientist, astrophysicist, nor a renowned astronomer or mathematician. However, I have spent years strategizing about our cosmos. Our universe has fascinated me. Now, I'm voicing my views. But, first let me ask you, 'If you had the ambition to write a new universe theory, how would you attempt it?' No doubt it would be a gargantuan task, thwart with enormous complexity and difficulty.

For starters, inside me there had always existed an innate need to better understand our lives. Like many people, my eyes looked to the sky, wondering what cosmos is all about.

Yet, it pleases me to live in a world where those more trained and smarter than me have worked hard for centuries to gain insight and offer cosmic explanations. Truth always emerges victorious, even if taking decades or centuries.

Through the mid 1970's and 80's every SF writer's work, along with science-astronomy magazines, passed through my hands *Cosmos,* by Carl Sagan, became my favorite TV show in 1980. At that early stage, I had no inkling to write new theory, as this was just a wonderful side line hobby. Nonetheless, my personal study of the cosmos had begun in earnest making me a self-proclaimed amateur cosmologist, gathering info about the origin, structure, dynamics, natural laws, and eventual fate of our universe. Soon, my note book grew fatter overflowing with new ideas, many disagreeing with the notions put forth by Big Bang theory.

I believe that my greatest forte was strategy. My strategic mind thought long and hard about the concepts read about in science and astronomy magazines. I began to envision the cosmos more from a strategic point of view, thinking about how a creator, artisan, or corporation might have built it all.

By 1997, I had begun writing in earnest, developing more ideas into my personal universe theory. I watched closely for new concepts that scientists and astronomers came up with, poised and ready to pounce by either accepting or rejecting how they saw our universe. Slowly, I was weaving a mosaic of my own theory placed into my logs. In my readings, it always seemed that Albert Einstein had been such a gigantic key factor in human history; unlocking wondrous universe mysteries. I bought into most of what Einstein had been selling; $E=mc^2$ was a most phenomenal discovery. Energy and matter had a close back and forth *forever* relationship.

In 1998, I began the finalization of the writing of my new theories into a book called Trillion Years Theory. But at that time, little did I know that life would get in the way and it would be 2013, 2014, and 2015 before my new theories were placed into print and published. Like good wine, time was well spent getting it right and perfected for consumption."

Interviewer: "Thanks for sharing your unique perspective of our cosmos. You've supplied plenty to ponder over. I'll never see our cosmos the same way after viewing your astounding theory. I look forward to more of your writings."

Author: "Thanks for reading. I'd like to congratulate you on reading the entire book. TT is not an easy read, because all the way through your mind is busy with comparisons to Big Bang and thoughts as to the validity of new Trillion Theory. Please keep up with my website where I'll update all the new books and projects which are planned with Trillion Theory.

So again, thank you for reading. Please keep up with my present website **www.trillionist.com** where there will also find a link to my future website *The Universe Trillion Project.*

Hopefully Trillion Theory will be your new theory too or at least open up for you some new thought corridors.

The future will definitely bring many new amazing cosmic discoveries. In 2018, NASA launches the James Webb Space Telescope as its new premiere space observatory to further study the formation of stars and planets and to study the birth and evolution of galaxies. Also, to hopefully discover alien life on other Earth-like planets. Furthermore, it would be great to also discover new evidence to authenticate the theories within TT and thereby prove Trillion Theory."

UNIVERSE TRILLION CLUB

Each day, more new members add their name to the *Universe Trillion Club.* These people support my initiatives and endeavors to more closely examine our universe.

You can play a role.

To sign up, simply email to:

trillionist2@gmail.com

Add your name to the website list at:

www.trilllionist.com

Thanks once again for reading
'Black Holes Built Our Cosmos' theory.
Cheers to you from author Ed Lukowich.

The End

Universe books authored by Ed Lukowich: